Undergraduate Texts in Mathematics

Editors
J.H. Ewing
F.W. Gehring
P.R. Halmos

Undergraduate Texts in Mathematics

(continued after index)

L. Christine Kinsey

Topology
of Surfaces

With 276 Illustrations

1113394 5 514 KIN

L. Christine Kinsey
Department of Mathematics
Canisius College
Buffalo, NY 14208 USA

Editorial Board:

John H. Ewing
Department of Mathematics
Indiana University
Bloomington, IN 47405 USA

F.W. Gehring
Department of Mathematics
University of Michigan
Ann Arbor, MI 48109 USA

Paul R. Halmos
Department of Mathematics
Santa Clara University
Santa Clara, CA 95053 USA

Mathematics Subject Classifications (1991): 54-01, 55-01

Library of Congress Cataloging-in-Publication Data
Kinsey, L. Christine.
    Topology of surfaces / L. Christine Kinsey.
        p.   cm. — (Undergraduate texts in mathematics)
    Includes bibliographical references and index.
    ISBN 0-387-94102-9 (New York : acid-free). — ISBN 3-540-94102-9
(Berlin : acid-free)
    1. Topology.   I. Title.   II. Series.
QA611.K47   1993
514 — dc20   93-2605

Printed on acid-free paper.

Production managed by Henry Krell; manufacturing supervised by Vincent Scelta.
Photocomposed copy prepared from author's TeX files.
Printed and bound by Braun-Blumfield, Ann Arbor, MI.
Printed in the United States of America.

9 8 7 6 5 4 3 2 1

ISBN 0-387-94102-9 Springer-Verlag New York Berlin Heidelberg
ISBN 3-540-94102-9 Springer-Verlag Berlin Heidelberg New York

30/5/96.

Preface

"...that famous pedagogical method whereby one begins with the general and proceeds to the particular only after the student is too confused to understand even that anymore."

Michael Spivak

This text was written as an antidote to topology courses such as Spivak describes. It is meant to provide the student with an experience in geometric topology. Traditionally, the only topology an undergraduate might see is point-set topology at a fairly abstract level. The next course the average student would take would be a graduate course in algebraic topology, and such courses are commonly very homological in nature, providing quick access to current research, but not developing any intuition or geometric sense. I have tried in this text to provide the undergraduate with a pragmatic introduction to the field, including a sampling from point-set, geometric, and algebraic topology, and trying not to include anything that the student cannot immediately experience. The exercises are to be considered as an integral part of the text and, ideally, should be addressed when they are met, rather than at the end of a block of material. Many of them are quite easy and are intended to give the student practice working with the definitions and digesting the current topic before proceeding. The appendix provides a brief survey of the group theory needed. The choice of topics addressed is my own, and fairly random at that, and was dictated by the fact that the text is used in a semester course. I often augment the text with articles from the general scientific and undergraduate math journals, which is a gentle way of getting the student to read and discuss mathematics.

This book originated as a set of lecture notes for a course in geometric topology at Vanderbilt University, and later at Canisius College. The present text has been used for a course taught to senior math majors. I have also taught a similar course for sophomores and juniors who have had a course in linear algebra but not in group theory, in which case I covered Chapters 1, 2, 4, 5, 6, 7, and 10.

This text was prepared using the $\mathcal{A}\mathcal{M}\mathcal{S}$-TEX macro package and the macros supplied by Springer-Verlag New York. Portions of the text were prepared with the support of a Canisius College faculty fellowship. I would like to thank Metod Alif, Terry Bisson, Jim Catalano, Alex Douglas, Richard Escobales, Jim Huard, Mike Mihalik, Ray Mullican, John Ratcliffe, Tina Romance, Karl Schroeder, and Marlene Trzaska for many corrections, comments, and suggestions. Any mistakes remain my own.

L. Christine Kinsey
Canisius College

Contents

Chapter 1

Introduction to topology

"when George drew out a tin of pineapple from the bottom of the hamper ... we felt that life was worth living after all ... there was no tin-opener to be found ... I took the tin off myself and hammered at it till I was worn out and sick at heart, whereupon Harris took it in hand. We beat it out flat; we beat it back square; we battered it into every form known to geometry — but we could not make a hole in it. Then George went at it, and knocked it into a shape, so strange, so weird, so unearthly in its wild hideousness, that he got frightened."

Jerome K. Jerome

1.1 An overview

The first thing we must do is firmly put down any notion you may have that we're about to start making maps. The only thing topology and topography have in common is their derivation from the Greek τοποσ, or space. Topology is a relatively new field of mathematics and is related to geometry. In both of these subjects one studies the shape of things. In geometry, one characterizes, for example, a can of pineapple by its height, radius, surface area, and volume. In topology, one tries to identify the more subtle property that makes it impossible to get the pineapple out of the tin, no matter what shape it is battered into, as long as one does not puncture the can.

In a geometry course, one studies at first simple figures such as triangles or quadrilaterals and their properties, such as the lengths of the sides, the measures of the angles, and the areas enclosed. Next, one develops a definition of when figures are geometrically the same or *congruent*. Congruent figures will have equal geometric properties: lengths, angle measures, areas, etc. One can study the set of functions, called isometries or rigid motions (such as rotations), which preserve congruence (so that $f(\Delta)$ is congruent to Δ). One develops theorems to tell when figures are congruent (side-angle-side congruence, etc.). One identifies the properties of certain classes of figures, such as the angle-sum theorem for triangles or the Pythagorean Theorem for right triangles.

We wish to emulate this course of study for topology and the more elusive properties not detected by geometry. Since length and angle measure are adequately covered in geometry, one decides to ignore these factors. Any two line segments, even of different lengths, are considered to be topologically equal, since by stretching one could be turned into the other. All angles are equal, since one could be bent to form the other. Thus, a square is the same topological shape as a rectangle or any other quadrilateral. By straightening one of the angles, a rectangle is seen to be the same as a triangle. One is also allowed to flex or bend lines into squiggles or curves. Thus, in topology, one considers the examples in each row of Figure 1.1 below to be same in topology, since they differ only in the geometric properties of length, angle measure, and curvature.

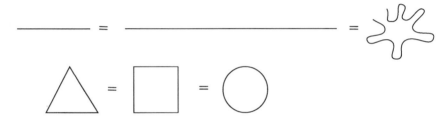

Fig. 1.1. The figures in each row are topologically the same

One concentrates, instead, on how a line segment and a circle differ. It seems obvious that the line and circle differ in some way not explicitly described by length or curvature or any other classical geometric property. The essence of this difference is the property that the circle or ring divides the plane into a region inside the circle and another region outside, but the line segment does not divide the plane. Even an infinite line, which does divide the plane, does so in a manner intrinsically different from the circle. We wish to quantify those properties in which line and the circle and similar figures differ and to define when objects are to be considered topologically equal and what properties are preserved by this type of equality.

Two objects are topologically identical if there is a continuous deformation (to be more precisely defined later) from one to the other. The bending and stretching that we allowed in Figure 1.1 are examples of such continuous deformations. Thus, topology is sometimes called the study of continuity, or, a more hackneyed term, "rubber-sheet geometry," as one pretends everything is formed of extremely flexible rubber. A mathematical cliché (included only in the interest of cultural literacy) is illustrated in Figure 1.2.

Fig. 1.2. Turning a doughnut into a coffee cup

To return to the analogy with the development of geometry, note that a crucial step is the definition and study of congruence. Congruence provides a notion of geometric equality: a sort of local equality relevant in a geometry class, but not at all appropriate in courses in linear algebra or calculus or topology. Congruent figures will have the same geometric properties: lengths of corresponding sides, angle measures, area, volume, perimeter, and curvature. Congruence is an example of a common construction in mathematics.

(1.1) Definition. *An equivalence relation, $\sim$, on a set of objects is a relation on the set such that:*

(1) *For each x in the set, $x \sim x$. [reflexivity]*
(2) *If $x \sim y$, then $y \sim x$. [symmetry]*
(3) *If $x \sim y$ and $y \sim z$, then $x \sim z$. [transitivity]*

An equivalence relation is a relation between objects that acts like equality. For example, if one is intimately acquainted with triangle A, and also knows that triangle B is congruent to A, as in Fig. 1.3, then one knows quite a bit about B: the lengths of its sides, the measures of the angles, the area, and perimeter.

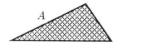
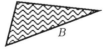

Fig. 1.3. Congruent triangles

These triangles are not equal, however, since they consist of different points. Congruence, a geometric equivalence relation, says nothing about the positioning of the figures, or their color, or smell, or what they taste like. Congruence only carries information about the geometric properties. Another geometric equivalence relation is similarity. Similar figures will have equal angle measures and proportional sides and areas but will not have equal lengths. Similarity picks up some, but not all, geometric data. The appropriate equivalence depends on the context and the purpose intended. Different fields in mathematics usually require different equivalences, defined for the objects in which one is currently interested and the properties one wishes to preserve. We wish to define and study a notion of topological equality that will determine when objects are the same topologically, though they may differ geometrically or in other ways.

(1.2) Definition. *Let A and B be topological spaces. Then A is topologically equivalent or homeomorphic to B if there is a continuous invertible function* $f : A \to B$ *with continuous inverse* $f^{-1} : B \to A$*. Such a function f is called a homeomorphism.*

This definition will be discussed in more detail in the next chapter, where it will also be shown that this is an equivalence relation, but one can assume that you have some familiarity with the concepts of continuity and invertibility from previous courses. An operation that is not continuous is cutting a figure in two. The inverse of cutting is pasting, which reverses the cutting operation, and thus is also not a homeomorphism. Consider the examples of Fig. 1.4.

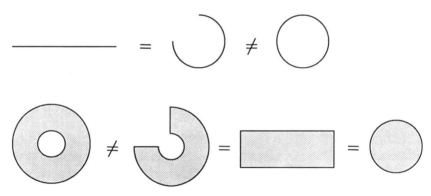

Fig. 1.4. More examples

Try the following exercise, using your intuitive idea of topological equivalence:

▷ **Exercise 1.1.** Classify up to topological equivalence, i.e., sort by topological types:

ABCDEFGHIJKLMNOPQRSTUVWXYZ0123456789

One also wishes to identify topological properties, just as in geometry one studies the properties of length, angle measure, area, volume, and curvature. A topological property is one which all topologically equivalent spaces share. Thus, if one knows that object A has topological properties X, Y, and Z, and that B is homeomorphic to A, then B must also have properties X, Y, and Z. Also, if C does not have property X, it follows that C cannot be homeomorphic to A or B. This text will pursue the search for topological properties through the three main subfields of topology in order to see a bit of each.

Chapters 2 and 3 are devoted to an introduction to *point-set topology*. One can think of this as the study of topological objects under a microscope, looking for properties that are decided at the level of points. An example of such a property is *connectedness*, which means just want one thinks it does, though, as is common in mathematics, writing down a formal definition is another matter entirely. Thus, the figure on the left in Figure 1.5 is connected, but removing a single point gives the figure on the right, which is not connected. Therefore, these figures cannot be homeomorphic.

Fig. 1.5. The figure on the left is connected, whereas that on the right is not connected

Geometric topology takes a more macroscopic view, looking for more global properties. Such a property is that of enclosing a cavity or hollow, one of the essential differences between a sphere and a disc.

Fig. 1.6. Sphere and disc

The particular approach to geometric topology we will take is *combinatorial*, in which the objects under study will be divided into nice pieces. This is a common mode of attack in mathematics — breaking a problem up into nice, manageable, predictable components, and then studying the components and the way in which they are put together. Chapters 4 and 5 are devoted to an introduction to the problems and techniques of combinatorial topology.

Algebraic topology is the search for something computable to identify topological properties, such as something that will say $\alpha(X) = 0$ if X is connected, but $\alpha(X) = 1$ if X is not connected, or $\beta(X) = 1$ if X encloses a cavity but $\beta(X) = 0$ if X does not. In keeping with the subtlety of the properties with which we are dealing, numbers alone will not usually suffice, and so it is necessary to use the more sophisticated group theory. The requisite group theory is developed in the Appendix; just enough for our purposes. Homology theory, one of the two main types of algebraic topology, is developed in Chapters 6 and 7, and homotopy theory is briefly described in Chapter 9.

Chapter 11 discusses ways in which topology affects subjects you have already studied; specifically, calculus and differential equations.

A natural question is, of course, why bother? Aside from parlor tricks, such as the doughnut-into-coffee cup routine mentioned earlier, and removing one's vest without taking off one's jacket (borrow clothes at least two sizes too large before attempting this), a number of exciting and useful applications will be proven, mostly in Chapter 10.

Theorem. *Everyone has at least one whirl or cowlick on his head.*

Theorem. *At any moment there are two locations on the equator exactly opposite each other with the same temperature.*

Theorem. *One needs no more than seven colors to color any map drawn on the surface of a doughnut so that no two adjacent areas have the same color.*

Theorem. *Given a sandwich consisting of a mass of bread, a mass of cheese, and a mass of ham (each mass may consist of several parts), it is possible to divide the sandwich in two by a single knife slice so that each part has half of each component.*

Theorem. *One cannot comb the hair on a coconut.*

Theorem. *At any time, there is a point on the earth where the wind is not blowing.*

Chapter 2

Point-set topology in $\mathbb{R}^n$

"Strange as it may sound, the power of mathematics rests in its evasion of all unnecessary thought and on its wonderful saving of mental operations."

Ernst Mach

In any mathematical study, the first thing to do is specify the type or category of objects to be investigated. In topology, the most general possible objects of study are sets of points with just enough structure to be able to define continuous functions. This, however, leads to a far more abstract study than is our purpose. We will, in this chapter, restrict our attention to subsets of real n-space, $\mathbb{R}^n$, with all the natural structure to which we are accustomed. This provides a class of objects with structure simple enough to be amenable to study while sufficiently rich in possibilities. Point-set topology is rather like looking at these sets of points under a microscope and identifying characteristics visible in that context. As in the beginning of any study, a great number of definitions, notation, and examples must be absorbed before proceeding. This pattern is rather like that commonly followed in calculus classes, where it is necessary to thoroughly understand limits and local behavior before proceeding to differentiation and integration. By basing everything firmly on the structure of euclidean space, many of the intricacies of general point-set topology can be avoided, but I have tried not to go so far in that direction as to give proper mathematicians the horrors. A number of simple proofs are included in the exercises.

2.1 Open and closed sets in $\mathbb{R}^n$

All of the point sets in this chapter will be subsets of real euclidean n-space, $\mathbb{R}^n = \{\mathbf{x} = (x_1, x_2, \ldots, x_n) : x_i \text{ a real number}\}$ where $\mathbf{x}$ denotes a point with n coordinates. For example, $\mathbb{R}^1$ or $\mathbb{R}$ is the real line, $\mathbb{R}^2 = \{\mathbf{x} = (x, y)\}$ is the standard euclidean plane, and $\mathbb{R}^3 = \{\mathbf{x} = (x, y, z)\}$ is 3-dimensional space. Generally, one uses x to denote a real number and $\mathbf{x}$ to denote a point in $\mathbb{R}^n$ for $n > 1$.

Given two points $\mathbf{x} = (x_1, x_2, \ldots, x_n)$ and $\mathbf{y} = (y_1, y_2, \ldots, y_n)$ in $\mathbb{R}^n$, the *euclidean distance* between $\mathbf{x}$ and $\mathbf{y}$ is measured by

$$\|\mathbf{x} - \mathbf{y}\| = \sqrt{(x_1 - y_1)^2 + (x_2 - y_2)^2 + \cdots + (x_n - y_n)^2}$$

The disc or ball centered at $\mathbf{x}$ with radius r in whichever space we are talking about is denoted by $D^n(\mathbf{x}, r)$. Formally,

$$D^n(\mathbf{x}, r) = \{\mathbf{y} \in \mathbb{R}^n : \|\mathbf{x} - \mathbf{y}\| < r\}$$

We make use of the standard conventions whereby (a, b) denotes the open interval $(a, b) = \{x \in \mathbb{R} : a < x < b\}$ and $[a, b]$ denotes the closed interval $[a, b] = \{x \in \mathbb{R} : a \le x \le b\}$. If the boundary or perimeter of a set pictured in the plane is drawn using a solid black curve, then those points on the boundary are to be considered to be included in the set. If drawn with a dashed or grey curve, the points on the edge are excluded.

In $\mathbb{R}^1$, $D^1(x, r)$ is an open interval centered at x (Fig. 2.1).

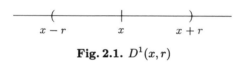

Fig. 2.1. $D^1(x, r)$

In $\mathbb{R}^2$, $D^2(\mathbf{x}, r)$ is a disc (not including the perimeter) with center $\mathbf{x}$ and radius r (Fig. 2.2).

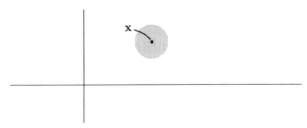

Fig. 2.2. $D^2(\mathbf{x}, r)$

In $\mathbb{R}^3$, $D^3(\mathbf{x}, r)$ is a solid ball (not including the bounding sphere) with center $\mathbf{x}$ and radius r.

(2.1) Definition. *These discs, $D^n(\mathbf{x}, r)$, are called (disc) neighborhoods of $\mathbf{x}$ in $\mathbb{R}^n$.*

For arbitrary sets of points in $\mathbb{R}^n$, we define the various ways a point **x** may lie in relationship to the set A. Let $\mathbb{R}^n - A$ designate all points not in A; this is called the *complement* of A. Of course, any point **x** is either in A or in $\mathbb{R}^n - A$, but we want to distinguish if **x** is buried in the middle of A, or on its fringes, or well outside of A.

(2.2) Definition. *Let A be a set of points with $A \subseteq \mathbb{R}^n$.*

   (1) *A point **x** is an interior point of A if there is a neighborhood N of **x** so that*

$$\mathbf{x} \in N \subseteq A$$

   i.e., the disc is totally enclosed in A.

   (2) *A point **x** is an exterior point of A if there is a neighborhood N so that*

$$\mathbf{x} \in N \subseteq \mathbb{R}^n - A$$

   i.e., the disc is totally outside of A. Another way of saying this is $N \cap A = \emptyset$.

   (3) *A point **x** is a limit point of A if every neighborhood N of **x** contains at least one point of A. Thus,*

$$N \cap A \neq \emptyset$$

*Note that every point of A is by this definition a limit point of A. If **x** itself is in A and is the only point in A which lies in a neighborhood N (i.e., $N \cap A = \{\mathbf{x}\}$), then **x** is called an isolated point of A. If **x** is a limit point of A such that every neighborhood of **x** contains a point $\mathbf{y} \notin A$, then **x** is called a frontier point of A.*

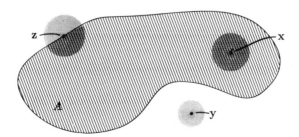

Fig. 2.3. Point **x** is interior to A, **y** is exterior, and **z** is a frontier point of A

Note that the frontier points are also limit points by the definition. If a point $\mathbf{x}$ is a limit point but not a frontier point, then the statement "every neighborhood of $\mathbf{x}$ contains a point $\mathbf{y} \notin A$" must be false, so "there is a neighborhood of $\mathbf{x}$ which contains no points not in A" must be true; thus, "there is a neighborhood of $\mathbf{x}$ which contains only points of A" is true, so $\mathbf{x}$ is an interior point of A. Therefore,

$$\{\text{limit points of } A\} = \{\text{frontier points of } A\} \cup \{\text{interior points of } A\}$$

The three options of interior, exterior, and frontier are both inclusive (i.e., every point in $\mathbb{R}^n$ must be either interior, exterior, or a frontier point of A) and mutually exclusive (i.e., a point cannot be both interior and exterior to A, etc.). Also, note that interior points are elements of A and exterior points are in the complement of A, whereas limit points and frontier points may lie in A or in its complement.

One way of picturing the definition above is to imagine that one is looking through a microscope. As one increases the magnification, the field of view includes points closer and closer to $\mathbf{x}$ (Fig. 2.4).

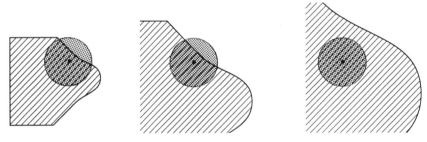

Fig. 2.4. The point $\mathbf{x}$ is interior to the shaded set A

The point $\mathbf{x}$ is interior to A if, under high enough magnification, the field of view contains only points of A, exterior if only points outside of A, and a frontier point if, under all possible changes of magnification, the field of view contains both points of A and points of $\mathbb{R}^n - A$.

(2.3) Definition. *A set A is open if every $\mathbf{x} \in A$ is an interior point.*

(2.4) Definition. *A set A is closed if every $\mathbf{x} \notin A$ is an exterior point.*

Many sets are neither open nor closed; for example, let A be a disc including the left semicircle, but not the right, as in Figure 2.5. In this example, $\mathbf{x}$ is a point in A that is not an interior point, so A cannot be open. The point $\mathbf{y}$ is not in A but is also not exterior to A, so A is not closed.

Fig. 2.5. A set which is neither open nor closed

(2.5) Definition. *A set $A \subseteq \mathbb{R}^n$ is bounded if $A \subseteq D^n(0, r)$ for some r. Thus, the set A can be enclosed in some sufficiently large disc.*

Bounded sets seem, in a sense, finite, since they do not go on forever. However, it will later appear that this is not the correct notion for topological finiteness.

▷ **Exercise 2.1.** Determine whether the following subsets of $\mathbb{R}^2$ are open, closed, and/or bounded.

   (1) $A = \{\|\mathbf{x}\| \le 1\}$
   (2) $B = \{\|\mathbf{x}\| = 1\}$
   (3) $C = \{\|\mathbf{x}\| < 1\}$
   (4) $D = \{\text{the } x\text{-axis}\}$
   (5) $E = \mathbb{R}^2 - \{\text{the } x\text{-axis}\}$
   (6) $F = \{(x, y) : x \text{ and } y \text{ are integers}\}$
   (7) $G = \{(1, 0), (1/2, 0), (1/3, 0), (1/4, 0), \ldots, (1/n, 0), \ldots\}$
   (8) $H = \mathbb{R}^2$
   (9) $I = \emptyset$, the null set.

(2.6) Definition. *The interior of a set A, denoted $Int(A)$, is the set of all interior points of A.*

Note that $Int(A) \subseteq A$ for any set A, but some sets have no interior. For example, let A consist of a single point $\mathbf{x}$, then

$$Int(A) = \emptyset \subseteq A = \{\mathbf{x}\}.$$

(2.7) Definition. *The frontier of a set A, denoted $Fr(A)$, is the set of all frontier points of A.*

Frontier points may be either in A or not, as in the example in Fig. 2.5, where both $\mathbf{x} \in A$ and $\mathbf{y} \notin A$ are frontier points.

(2.8) Definition. *The closure of A, denoted $Cl(A)$, is the set $A \cup Fr(A)$.*

Note that $A \subseteq Cl(A)$ for any set A.

▷ **Exercise 2.2.** Show that $Cl(A) = \{$limit points of A$\}$.

▷ **Exercise 2.3.** Find the interiors, frontiers, and closures of the sets in Exercise 2.1 above.

As an example of the type of argument used in the next set of exercises, we prove (with commentary in smaller type):

(2.9) Theorem. *For any set A in $\mathbb{R}^n$, the set $Cl(A)$ is a closed set.*

Proof. By Definition 2.8, $Cl(A) = A \cup Fr(A)$. We must show that $Cl(A)$ is closed. By Definition 2.4, this means we must show that if $\mathbf{x}$ is an arbitrary point with $\mathbf{x} \notin Cl(A)$, then $\mathbf{x}$ is exterior to $Cl(A)$. This can be restated using Definition 2.2: If $\mathbf{x} \notin Cl(A)$, then there is a disc $D^n(\mathbf{x}, r)$ such that $D^n(\mathbf{x}, r) \cap Cl(A) = \emptyset$. Since $\mathbf{x} \notin Cl(A)$ and $Cl(A) = A \cup Fr(A)$, $\mathbf{x}$ cannot be interior to A (since $Int(A) \subseteq A \subseteq Cl(A)$) nor can $\mathbf{x}$ be a limit point of A (since $Fr(A) \subseteq Cl(A)$), so $\mathbf{x}$ must be exterior to A. Thus, there is a disc $D^n(\mathbf{x}, r)$ with $D^n(\mathbf{x}, r) \cap A = \emptyset$. Does this mean $D^n(\mathbf{x}, r) \cap Cl(A) = \emptyset$ (and so $\mathbf{x}$ is exterior to $Cl(A)$ as well as A)?

Here we have recalled the definitions involved in the theorem ($Cl(A)$ and what it means for $Cl(A)$ to be closed) and then restated everything on the common ground of these definitions (exterior points and intersections of discs with the sets involved).

Let us try a new technique: proof by contradiction. Assume everything in the hypothesis is true, but for some reason the conclusion of the theorem (that $Cl(A)$ is closed) is false.

If $Cl(A)$ is not closed, then by Definition 2.4 we have a point $\mathbf{x}$ which is not in $Cl(A)$ but is not exterior to $Cl(A)$. From the discussion above, $\mathbf{x} \notin Cl(A)$ implies that there is a disc $D^n(\mathbf{x}, r) \cap A = \emptyset$, but since $\mathbf{x}$ is not exterior to $Cl(A)$, the disc $D^n(\mathbf{x}, r)$ must overlap $Cl(A)$. So there has to be a point $\mathbf{y} \in D^n(\mathbf{x}, r) \cap Cl(A)$. The point $\mathbf{y} \notin A$ because $D^n(\mathbf{x}, r) \cap A = \emptyset$. Since $\mathbf{y} \in Cl(A) = A \cup Fr(A)$, $\mathbf{y}$ must be a frontier point of A and every disc around $\mathbf{y}$ must contain points of A.

Draw a picture of $\mathbf{x}$ and $\mathbf{y}$ and the disc to help visualize the relationships. See Figure 2.6.

Let $r' = r - \|\mathbf{x} - \mathbf{y}\|$, and note that $D^n(\mathbf{y}, r') \subseteq D^n(\mathbf{x}, r)$. Then $D^n(\mathbf{y}, r')$ is a disc about $\mathbf{y}$ and must contain points of A by the comments above. However, $D^n(\mathbf{y}, r')$ is inside $D^n(\mathbf{x}, r)$, so $D^n(\mathbf{x}, r)$ contains points of A.

This contradicts the assumptions we have made.

Thus, $\mathbf{y}$ cannot exist and so $D^n(\mathbf{x}, r) \cap Cl(A) = \emptyset$. Thus, $\mathbf{x}$ must be exterior to $Cl(A)$. Therefore, $Cl(A)$ is closed.   □

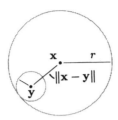

Fig. 2.6. $D^n(\mathbf{y}, r') \subseteq D^n(\mathbf{x}, r)$

Proof by contradiction or *reductio ad absurdum* is a time-honored technique used on theorems of the form "If A is true, then B is true." One assumes that A is true but that B is false and proceeds in this line of logic until coming to a patent absurdity: something like "0=1", or "x is both here and there." One is then forced to concede that something is wrong. If A really is true (and one has not made a mistake in the course of the argument), then B must be true, rather than false. This is all rather like ignoring reality until one runs one's head into a brick wall. It is a technique I tend to employ often, so you will be seeing it again. This is, perhaps, a temperamental idiosyncrasy. It is a grand mathematical tool, but a hell of a way to run one's life.

▷ **Exercise 2.4.** Prove that a set A is closed if and only if $Fr(A) \subseteq A$.

▷ **Exercise 2.5.** A set A is closed if and only if $A = Cl(A)$.

▷ **Exercise 2.6.** For any set A, $Fr(A)$ is closed.

▷ **Exercise 2.7.** For any set $A \subseteq \mathbb{R}^n$, $Fr(A) = Fr(\mathbb{R}^n - A)$.

▷ **Exercise 2.8.** For any set $A \subseteq \mathbb{R}^n$, $Fr(A) = Cl(A) \cap Cl(\mathbb{R}^n - A)$.

▷ **Exercise 2.9.** A set A is open if and only if $A = Int(A)$.

▷ **Exercise 2.10.** A set $A \subseteq \mathbb{R}^n$ is open if and only if $\mathbb{R}^n - A$ is closed.

▷ **Exercise 2.11.** A set $A \subseteq \mathbb{R}^n$ is closed if and only if $\mathbb{R}^n - A$ is open.

▷ **Exercise 2.12.** If A and B are any sets, prove that

$$Cl(A \cap B) \subseteq Cl(A) \cap Cl(B)$$

Give an example where $Cl(A \cap B) = \emptyset$, but $Cl(A) \cap Cl(B) \neq \emptyset$.

▷ **Exercise 2.13.** If A and B are open sets, prove that $A \cup B$ and $A \cap B$ are open. Give an example of a sequence of open sets $A_1, A_2, A_3, \ldots$ such that $\bigcap_{n=1}^{\infty} A_n$ is not open.

▷ **Exercise 2.14.** If $\{A_\alpha\}$ is a collection of open sets, prove that $\bigcup_\alpha A_\alpha$ is an open set.

▷ **Exercise 2.15.** If A and B are closed sets, prove that $A \cup B$ and $A \cap B$ are closed. Give an example of a sequence of closed sets $B_1, B_2, B_3, \ldots$ such that $\bigcup_{n=1}^{\infty} B_n$ is not closed.

▷ **Exercise 2.16.** If $\{A_\alpha\}$ is a collection of closed sets, prove that $\bigcap_\alpha A_\alpha$ is a closed set.

Most of these are simple proofs, involving only the unraveling of the appropriate definitions (as in the first paragraph of Theorem 2.9).

Note that Exercises 2.13 and 2.14 imply that any union of open sets is open, and any finite intersection of open sets is open, but an infinite intersection of open sets need not be open. Similarly, Exercises 2.15 and 2.16 show that any intersection of closed sets is closed, and any finite union of closed sets is closed, although an infinite union of closed sets may not be closed.

A *sequence* in A is an infinite ordered set of points

$$\{\mathbf{x}_i\}_{i=1}^{\infty} = \{\mathbf{x}_1, \mathbf{x}_2, \mathbf{x}_3, \ldots\}$$

where $\mathbf{x}_i \in A \subseteq \mathbb{R}^n$.

(2.10) Definition. *A point $\mathbf{x}$ is a limit point of a sequence $\{\mathbf{x}_i\}_{i=1}^{\infty}$ if every neighborhood of $\mathbf{x}$ contains an infinite number of the $\mathbf{x}_i$'s.*

The difference between a sequence and a set is illustrated by the sequence $\{(-1)^k\}_{k=1}^{\infty}$ in $\mathbb{R}$. As a set, this consists of two elements $\{-1, 1\} = B$, but as a sequence, this is the infinite list $\{-1, 1, -1, 1, -1, 1, \ldots\}$. This sequence has two limit points: -1 and $+1$, both of which are also limit points of the set B. Another example is the sequence $\{1, \frac{1}{2}, 1, \frac{1}{3}, 1, \frac{1}{4}, 1, \frac{1}{5}, \ldots\}$. As a set, $\{1, \frac{1}{2}, 1, \frac{1}{3}, 1, \frac{1}{4}, \ldots\}$ is the same as $C = \{1, \frac{1}{2}, \frac{1}{3}, \frac{1}{4}, \ldots\}$, but the sequence has two limit points: 1 since it is repeated infinitely many times, and 0 as the limit of the sequence $\{1/n\}$. Any neighborhood of 0 will be of the form $D^1(0, r) = \{x \in \mathbb{R} : |x| < r\}$ for some radius r. Choose some (large) integer N so that $\frac{1}{N} < r$ (or $N > \frac{1}{r}$). Any $n \geq N$ will have $\frac{1}{n} \leq \frac{1}{N} < r$, so $\frac{1}{n} \in D^1(0, r)$ for $n \geq N$. Thus, any neighborhood of 0

contains infinitely many elements of the sequence $\{1, \frac{1}{2}, 1, \frac{1}{3}, 1, \frac{1}{4}, \dots\}$, so 0 is a limit point of the sequence. The set $C = \{1, \frac{1}{2}, \frac{1}{3}, \frac{1}{4}, \dots\}$ has limit points $\{1, \frac{1}{2}, \frac{1}{3}, \frac{1}{4}, \dots\} \cup \{0\}$, where $\{1, \frac{1}{2}, \frac{1}{3}, \frac{1}{4}, \dots\}$ are limit points because any element of C is a limit point of C (recall the definition of a limit point) and 0 is a limit point since any neighborhood of 0 contains points of the set C. The following theorem shows that a limit point of a set may also be considered as a limit point of some sequence of points chosen from the set.

(2.11) Theorem. *If $\mathbf{x}$ is a limit point of a set A in $\mathbb{R}^n$, then there is a sequence of points $\{\mathbf{x}_i\}_{i=1}^{\infty}$ where $\mathbf{x}_i \in A$ so that $\mathbf{x}$ is a limit point of the sequence $\{\mathbf{x}_i\}_{i=1}^{\infty}$.*

Proof. If $\mathbf{x}$ is a limit point of the set A, then every disc about $\mathbf{x}$ contains points of A. We wish to choose a sequence of points $\mathbf{x}_k$ in A with limit $\mathbf{x}$. By Definition 2.2, since $\mathbf{x}$ is a limit point of A, every disc neighborhood of $\mathbf{x}$ must contain some points of A. Consider the family of discs $D^n(\mathbf{x}, \frac{1}{k})$ for $k = 1, 2, 3, \dots$. These are discs centered at $\mathbf{x}$ with radii $1, \frac{1}{2}, \frac{1}{3}, \dots$, and must contain points of A. Choose $\mathbf{x}_k \in D^n(\mathbf{x}, \frac{1}{k}) \cap A$ for each k. This gives us a sequence of points from A. For any disc $D^n(\mathbf{x}, r)$, an integer N may be chosen so that $\frac{1}{N} < r$ (or $N > \frac{1}{r}$). It follows that the discs are nested inside each other: $D^n(\mathbf{x}, r) \supseteq D^n(\mathbf{x}, \frac{1}{N}) \supseteq D^n(\mathbf{x}, \frac{1}{N+1}), \dots$, and so

$$\mathbf{x}_N, \mathbf{x}_{N+1}, \mathbf{x}_{N+2}, \dots \in D^n\left(\mathbf{x}, \frac{1}{N}\right) \subseteq D^n(\mathbf{x}, r)$$

Therefore, $\mathbf{x}$ is a limit point of $\{\mathbf{x}_k\}_{k=1}^{\infty}$. □

▷ **Exercise 2.17.** Show that $\mathbf{x}$ is the only limit point of the sequence constructed in Theorem 2.11.

The converse of Theorem 2.11 is also true:

(2.12) Theorem. *If $\{\mathbf{x}_i\}_{i=1}^{\infty}$ is a sequence with each $\mathbf{x}_i \in A \subseteq \mathbb{R}^n$, and $\mathbf{x}$ is a limit point of the sequence $\{\mathbf{x}_i\}_{i=1}^{\infty}$, then $\mathbf{x}$ is a limit point of the set A.*

▷ **Exercise 2.18.** Prove Theorem 2.12.

These theorems show the connection between limit points for sets and limit points for sequences. Be careful, though, since if one again considers the example $\{1, \frac{1}{2}, 1, \frac{1}{3}, 1, \frac{1}{4}, 1, \frac{1}{5}, \dots, 1, \frac{1}{n}, \dots\}$, Theorem 2.12 says the points 0 and 1 are limit points of the set, since they are limit points of the sequence. On the other hand, $\frac{1}{4}$ is a limit point of the set, but Theorem 2.11 does not imply that $\frac{1}{4}$ is a limit point of the sequence $\{1, \frac{1}{2}, 1, \frac{1}{3}, 1, \frac{1}{4}, 1, \frac{1}{5}, \dots, 1, \frac{1}{n}, \dots\}$, but that another sequence can be constructed from this set with $\frac{1}{4}$ as limit point. For example, $\{\frac{1}{4}, \frac{1}{4}, \frac{1}{4}, \frac{1}{4}, \dots\}$ is a sequence in the set $\{1, \frac{1}{2}, \frac{1}{3}, \frac{1}{4}, \frac{1}{5}, \dots\}$ with the desired limit $\frac{1}{4}$.

2.2 Relative neighborhoods

There is a certain difficulty or ambiguity with the material above. For example, consider the interval $(-1, 1)$. Since $(-1, 1) = D^1(0, 1)$, this is open in $\mathbb{R}^1$ but is not open when considered as a subset of the plane, since any disc about a point of $(-1, 1)$ will overlap with the upper and lower half-planes:

Fig. 2.7. The interval $(-1, 1)$ is open in $\mathbb{R}^1$, but not open in $\mathbb{R}^2$

The real line $\mathbb{R}^1$ itself is both open and closed, but when considered as a subset of $\mathbb{R}^2$, it is not open. The definitions above depend on the context in which the set is viewed. Note that in Exercise 2.1, I was careful to state that the sets were viewed as subsets of $\mathbb{R}^2$. This section will remedy this problem by considering more carefully the context of the sets and neighborhoods.

(2.13) Definition. *Let $A \subseteq \mathbb{R}^n$. A neighborhood of a point $\mathbf{x} \in A$ relative to A is a set of the form $D^n(\mathbf{x}, r) \cap A$.*

For example, $\mathbb{R}^1 \subseteq \mathbb{R}^2$, so $\mathbb{R}^1$ has relative neighborhoods in $\mathbb{R}^2$ that look like discs intersected with $\mathbb{R}^1$. This gives exactly the same system of neighborhoods as the earlier definition, since a disc intersected with a line is an open interval. Thus, Definition 2.13 coincides with the earlier Definition 2.1 for this example. To see that this works in all dimensions, one chooses a point $\mathbf{x} \in \mathbb{R}^k$ and then one decides to raise the dimension of the surrounding space by enclosing $\mathbb{R}^k$ inside $\mathbb{R}^n$ for $n > k$, just as one includes the line $\mathbb{R}$ inside the plane $\mathbb{R}^2$, or the plane inside 3-space. In this situation, the point $\mathbf{x} = (x_1, x_2, \dots, x_k) \in \mathbb{R}^k$ becomes $\mathbf{x} = (x_1, x_2, \dots, x_k, 0, \dots, 0) \in \mathbb{R}^n$. A disc neighborhood of $\mathbf{x}$ in the larger space $\mathbb{R}^n$ intersected with the smaller space $\mathbb{R}^k$ will give

$$
\begin{aligned}
D^n(\mathbf{x}, r) \cap \mathbb{R}^k &= \{\mathbf{y} \in \mathbb{R}^n : \|\mathbf{x} - \mathbf{y}\| < r\} \cap \mathbb{R}^k \\
&= \{(y_1, y_2, \dots, y_k, \dots, y_n) : \\
&\quad \sqrt{(x_1 - y_1)^2 + \dots + (x_k - y_k)^2 + \dots + (0 - y_n)^2} < r\} \cap \mathbb{R}^k \\
&= \{(y_1, y_2, \dots, y_k) : \sqrt{(x_1 - y_1)^2 + \dots + (x_k - y_k)^2} < r\} \\
&= D^k(\mathbf{x}, r)
\end{aligned}
$$

Thus, the idea of relative neighborhoods gives the same sets regardless of the dimension of the surrounding euclidean space.

As another example, consider the interval $[-1, 1]$. The point $\mathbf{x} = 1$ has the set $(\frac{1}{2}, 1]$ as a neighborhood relative to $[-1, 1]$, since the intersection $D^1(1, \frac{1}{2}) \cap [-1, 1] = (\frac{1}{2}, 1]$. The set $(\frac{1}{2}, 1]$ of Fig. 2.8 is considered open in $[-1, 1]$, since every point in $(\frac{1}{2}, 1]$ has a neighborhood (relative to $[-1, 1]$) totally enclosed in $(\frac{1}{2}, 1]$. This set is clearly not open as a subset of $\mathbb{R}$.

$$-1 \qquad\qquad 0 \qquad \tfrac{1}{2} \quad 1$$

Fig. 2.8. $(\frac{1}{2}, 1]$ is open in $[-1, 1]$ but not in $\mathbb{R}$.

Another example is the cylinder, pictured as a subset of $\mathbb{R}^3$ in Fig. 2.9. The point $\mathbf{x}$ has for a neighborhood a slightly warped disc, the intersection of the cylinder with an open ball in $\mathbb{R}^3$, and $\mathbf{y}$ has for a neighborhood a warped half disc.

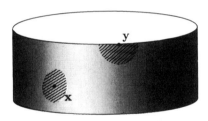

Fig. 2.9. The cylinder, with neighborhoods of $\mathbf{x}$ and $\mathbf{y}$

Recall that the definitions of interior, exterior, and limit points refer only to neighborhoods, not specifically to discs. These definitions remain valid for relative neighborhoods, as do the definitions, theorems, and exercises which follow. One need only keep clearly in mind the context space. For example, Definition 2.2 can be restated as:

(2.2′) Definition. *Let* $B \subseteq A$.

   (1) *A point* $\mathbf{x} \in A$ *is an interior point of* B *relative to* A *if there is a neighborhood* N *of* $\mathbf{x}$ *relative to* A *so that*

$$\mathbf{x} \in N \subseteq B$$

(2) *A point $\mathbf{x} \in A$ is an exterior point of B relative to A if there is a neighborhood N relative to A so that*

$$\mathbf{x} \in N \subseteq A - B$$

(3) *A point $\mathbf{x} \in A$ is a limit point of B relative to A if every neighborhood N relative to A has*

$$N \cap B \neq \emptyset$$

This, in turn, will change the definitions which rely on Definition 2.2'. For example:

(2.3′) Definition. *Let $B \subseteq A$. The set B is open relative to A if every $\mathbf{x} \in B$ is an interior point relative to A, i.e., if every $\mathbf{x} \in B$ has a relative neighborhood N with $\mathbf{x} \in N \subseteq B$.*

Exercise 2.13 can then be restated as: "If B and C are subsets of some set A, and B and C are open relative to A, then $B \cap C$ and $B \cup C$ are open relative to A."

The proofs of the theorems and exercises of Chapter 2.1 need only minor modifications (just intersect any discs with your background space) to remain valid, so we will not repeat them.

Another way of characterizing the relatively open and closed sets is the following theorem:

(2.14) Theorem. *If $B \subseteq A \subseteq \mathbb{R}^n$.*
  (1) *B is open relative to A if and only if $B = A \cap O$ for some set O which is open in $\mathbb{R}^n$.*
  (2) *B is closed relative to A if and only if $B = A \cap C$ for some set C which is closed in $\mathbb{R}^n$.*

Proof.
  (1) If B is open in A, then every point $\mathbf{x} \in B$ has a relative neighborhood $N_{\mathbf{x}}$ (with respect to A) with $\mathbf{x} \in N_{\mathbf{x}} \subseteq B$. By Definition 2.13, $N_{\mathbf{x}} = D^n(\mathbf{x}, r_{\mathbf{x}}) \cap A$ for some radius $r_{\mathbf{x}}$. Let $O = \bigcup_{\mathbf{x} \in B} D^n(\mathbf{x}, r_{\mathbf{x}})$. By Exercise 2.14, the set O is open in $\mathbb{R}^n$. Furthermore,

$$O \cap A = \left(\bigcup_{\mathbf{x} \in B} D^n(\mathbf{x}, r_{\mathbf{x}}) \right) \cap A$$
$$= \bigcup_{\mathbf{x} \in B} (D^n(\mathbf{x}, r_{\mathbf{x}}) \cap A)$$
$$= \bigcup_{\mathbf{x} \in B} N_{\mathbf{x}}$$

Clearly, $B \subseteq \bigcup_{\mathbf{x} \in B} N_{\mathbf{x}}$, since every $\mathbf{x} \in B$ is in one of the $N_{\mathbf{x}}$'s. Also, since every $N_{\mathbf{x}} \subseteq B$, it follows that $\bigcup_{\mathbf{x} \in B} N_{\mathbf{x}} \subseteq B$, so $B = \bigcup_{\mathbf{x} \in B} N_{\mathbf{x}} = O \cap A$.

Conversely, if $B = A \cap O$ for some set O which is open in $\mathbb{R}^n$, then every point $\mathbf{x} \in B$ is also in O. Since O is open, there is a disc $D^n(\mathbf{x}, r)$ with $\mathbf{x} \in D^n(\mathbf{x}, r) \subseteq O$. Let $N = D^n(\mathbf{x}, r) \cap A$. Then $\mathbf{x} \in N \subseteq O \cap A = B$, so B is open in A by Definition 2.3'.

(2) First note that the set B is closed relative to A if, for every point $\mathbf{x} \in A - B$, there is a relative neighborhood N with $N \cap B = \emptyset$ (this is the relative restatement of Definition 2.4). Thus, we have for each $\mathbf{x} \in A - B$ a neighborhood $N_{\mathbf{x}} = D^n(\mathbf{x}, r_{\mathbf{x}}) \cap A$ with $N_{\mathbf{x}} \cap B = \emptyset$. Let $O = \bigcup_{\mathbf{x} \in A - B} D^n(\mathbf{x}, r_{\mathbf{x}})$. As above, O is an open set. By Exercise 2.11, $C = \mathbb{R}^n - O$ will be closed. It is easy to see that $A - B = O \cap A$, and so $B = C \cap A$.

Conversely, if $B = C \cap A$ for some set C which is closed in $\mathbb{R}^n$, then every point $\mathbf{x} \in A - B$ will satisfy $\mathbf{x} \notin C$. Since C is closed, there is a disc $D^n(\mathbf{x}, r)$ with $\mathbf{x} \in D^n(\mathbf{x}, r)$ and $D^n(\mathbf{x}, r) \cap C = \emptyset$. Let $N = D^n(\mathbf{x}, r) \cap A$. Then $\mathbf{x} \in N$ and $N \cap B \subseteq N \cap C = \emptyset$. Thus, B is closed in A.

$\square$

▷ **Exercise 2.19.** Show that any set A is both open and closed relative to itself.

▷ **Exercise 2.20.** Show that $\emptyset$ is both open and closed relative to A for any set A.

▷ **Exercise 2.21.** Give an example of sets $B \subseteq A \subseteq \mathbb{R}^3$ where B is open relative to A but not open in $\mathbb{R}^3$.

▷ **Exercise 2.22.** Show that A is open relative to X if and only if $X - A$ is closed relative to X.

▷ **Exercise 2.23.** Show that A is closed relative to X if and only if $X - A$ is open relative to X.

2.3 Continuity

A *function* $f : D \longrightarrow R$ is a correspondence which associates to every point $\mathbf{x} \in D$ a unique point $f(\mathbf{x}) \in R$. The set D is called the *domain* of f and $R = f(D)$ the *range*. At this point of your training, many of you are so programmed that if one says to you "$f(x) = 2x$", the picture that springs to your mind is a line with slope 2 (Fig. 2.10).

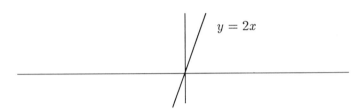

Fig. 2.10. The graph of $f(x) = 2x$

This is not at all what we want. Think again: if one is given a number, f gives you back a number twice as large: $f(0) = 0$, $f(1) = 2$, $f(-2) = -4$. If given the interval $[0, 1]$, f gives back the interval $[0, 2]$. The function f is stretching the real line by a factor of 2. You can feed in any real number and get back any real number, so $f : \mathbb{R} \longrightarrow \mathbb{R}$. The action of f can be pictured as in Figure 2.11.

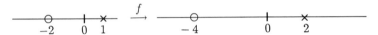

Fig. 2.11. The action of $f(x) = 2x$

Similarly, $g(x) = |x|$ folds the real line in half (Fig. 2.12).

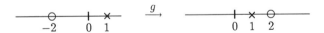

Fig. 2.12. The action of $g(x) = |x|$

Think of $h(x) = x + 3$ as shifting the real line 3 units to the right.

▷ **Exercise 2.24.** Discuss the action of the functions on $\mathbb{R}^2$ below, describing what each does to the points on the x-axis, the y-axis, the line $y = x$, and the unit square $\{(x, y) : 0 \le x \le 1, 0 \le y \le 1\}$

(1) $f(x, y) = (2x, 3y)$
(2) $g(x, y) = (-x, -y)$
(3) $h(x, y) = (y, x)$

Finally, we are ready to define and study continuity.

(2.15) Definition. *Let $D \subseteq \mathbb{R}^n$ and $R \subseteq \mathbb{R}^m$. A function $f : D \longrightarrow R$ is continuous if whenever A is an open set in R, then $f^{-1}(A)$ is an open set in D.*

Here f^{-1} denotes the set-theoretic inverse of f, so that f^{-1} is defined for any subset A in the range of f by $f^{-1}(A) = \{\mathbf{x} \in D : f(\mathbf{x}) \in A\}$. For example, if f is the function that folds the rectangle in half along the dotted line to give the picture in Figure 2.13, and A is the disc shown on the right, then $f^{-1}(A)$ consists of the two discs on the left. In general, f^{-1} is not a function. A point y (pictured below) on the fold has a half-disc neighborhood N relative to the rectangle, which is open relative to the rectangle on the right. Then $f^{-1}(N)$ is a disc, which is open in the rectangle on the left.

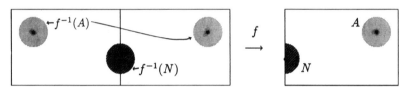

Fig. 2.13. The function f folds the strip in half

It is often easier to work with an alternate definition related to the one you may remember from a calculus course: f is continuous at x_0 if $\lim_{x \to x_0} f(x) = f(x_0) = f(\lim_{x \to x_0} x)$. The topological version of this is

(2.16) Theorem. *Let $D \subseteq \mathbb{R}^n$ and $R \subseteq \mathbb{R}^m$. A function $f : D \longrightarrow R$ is continuous if and only if whenever $\mathbf{x}$ is a limit point of a set B in D, then $f(\mathbf{x})$ is a limit point of the set $f(B)$ in R.*

Proof. First we will prove by contradiction that the limit point condition implies that the function is continuous. Assume that whenever $\mathbf{x}$ is a limit point of $B \subseteq D$, then $f(\mathbf{x})$ is a limit point of $f(B) \subseteq R$, but that there is an open set $A \subseteq R$ with $f^{-1}(A)$ *not* open so that Definition 2.15 is not satisfied. If $f^{-1}(A)$ is not open, then there is a point $\mathbf{x} \in f^{-1}(A)$ that is not interior to $f^{-1}(A)$. Thus, every neighborhood $D^n(\mathbf{x}, r)$ contains some point $\mathbf{x}_r \notin f^{-1}(A)$ (otherwise, we would have a neighborhood totally in $f^{-1}(A)$, and $\mathbf{x}$ would be an interior point). For the sequence of neighborhoods $D^n(\mathbf{x}, \frac{1}{k})$, choose points $\mathbf{x}_k \in D^n(\mathbf{x}, \frac{1}{k})$ with $\mathbf{x}_k \notin f^{-1}(A)$. The sequence (or set) $\{\mathbf{x}_k\}_{k=1}^{\infty}$ has limit point $\mathbf{x}$, so by the assumptions the sequence $f(\{\mathbf{x}_k\}_{k=1}^{\infty}) = \{f(\mathbf{x}_k)\}_{k=1}^{\infty}$ has limit point $f(\mathbf{x})$. Since $\mathbf{x}_k \notin f^{-1}(A)$, it follows that $f(\mathbf{x}_k) \notin A$. Thus, there are points not in A arbitrarily close to the point $f(\mathbf{x})$, so $f(\mathbf{x})$ is not an interior point. Therefore, A contains a non-interior point, so A cannot be open. This contradicts the choice of A, so something must be wrong with our assumptions: $f^{-1}(A)$ must be open, so f must be continuous.

Next, assume f is continuous as in Definition 2.15, but there is some set B in D with limit point $\mathbf{x}$ and $f(\mathbf{x})$ is *not* a limit point of the set $f(B)$. If $f(\mathbf{x})$ is not a limit point, it must be exterior to $f(B)$. Thus, there is a neighborhood N of $f(\mathbf{x})$ so that $N \cap f(B) = \emptyset$. Note that $\mathbf{x} \in f^{-1}(N)$ and, since f is continuous, $f^{-1}(N)$ is open. Thus, $\mathbf{x}$ has a neighborhood N' enclosed in $f^{-1}(N)$. Since $\mathbf{x}$ is a limit point of B, $\emptyset \neq N' \cap B \subseteq f^{-1}(N)$. Let $\mathbf{y} \in N' \cap B$. Then $f(\mathbf{y}) \in N$ and $f(\mathbf{y}) \in f(B)$ since $\mathbf{y} \in B$. This contradicts the assumption that $f(B) \cap N = \emptyset$. $\square$

Yet another equivalent definition of continuity is

(2.17) Theorem. *Let $D \subseteq \mathbb{R}^n$ and $R \subseteq \mathbb{R}^m$. A function $f : D \longrightarrow R$ is continuous at a point $\mathbf{x} \in D$ if and only if for any $\epsilon > 0$, there exists a $\delta > 0$ such that whenever $\mathbf{y} \in D$ and $\|\mathbf{x} - \mathbf{y}\| < \delta$, then $\|f(\mathbf{x}) - f(\mathbf{y})\| < \epsilon$.*

Proof. Let $\mathbf{x} \in D \subseteq \mathbb{R}^n$ and assume that f is continuous at $\mathbf{x}$ by Definition 2.15. Given an $\epsilon > 0$, consider the set $A = R \cap D^m(f(\mathbf{x}), \epsilon)$. Since this set is open in R, the inverse image $f^{-1}(A)$ is open in D and contains the point $\mathbf{x}$. Therefore, there is a relative neighborhood of $\mathbf{x}$ of the form $D^n(\mathbf{x}, \delta) \cap D \subseteq f^{-1}(A)$. Note that every point $\mathbf{y} \in D^n(\mathbf{x}, \delta) \cap D$ satisfies $\mathbf{y} \in D$ and $\|\mathbf{x} - \mathbf{y}\| < \delta$, and, since $f(\mathbf{y}) \in A$, $\|f(\mathbf{x}) - f(\mathbf{y})\| < \epsilon$.

If f satisfies the conditions of this theorem for every $\mathbf{x} \in D$, then we must show that f is continuous by Definition 2.15. Let A be an open set relative to R. We must show that $f^{-1}(A)$ is open relative to D. Let $\mathbf{x} \in f^{-1}(A)$. Then $f(\mathbf{x}) \in A$ and since A is open in R, there is a neighborhood of the form $D^m(f(\mathbf{x}), \epsilon)$ such that $D^m(f(\mathbf{x}), \epsilon) \cap R \subseteq A$. By the hypotheses, for this ϵ there exists a δ such that for all points $\mathbf{y} \in D \cap D^n(\mathbf{x}, \delta)$, $f(\mathbf{y}) \in D^m(f(\mathbf{x}), \epsilon) \cap R \subseteq A$. Therefore, $D \cap D^n(\mathbf{x}, \delta) \subseteq f^{-1}(A)$. Thus, $f^{-1}(A)$ is open in D. $\square$

The version of continuity of Theorem 2.17 is used in many analysis texts, but not commonly in topology. When dealing with functions of real numbers, you may assume that the usual addition, difference, multiplication, and division (except when dividing by zero) of continuous functions is continuous.

As an example, consider the function that takes the rectangle pictured in Figure 2.14 and cuts it in half along the dotted line. The points along the line must go somewhere, so we will put them in the left half, $f(A)$ (this is an arbitrary choice — put them anywhere you like). The point $\mathbf{x}$ on the dotted line in D is a limit point of the right side, B, so as in Theorem 2.11

we can choose a sequence $\{x_i\}_{i=1}^{\infty}$ of points in B with limit point x. After cutting, we get a sequence $\{f(x_i)\}_{i=1}^{\infty}$ of points in $f(B)$ with limit point along the missing edge of $f(B)$, but $f(x)$ is in $f(A)$. Thus, $f(x)$ cannot be a limit point of $\{f(x_i)\}_{i=1}^{\infty}$, so this function is not continuous.

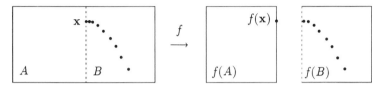

Fig. 2.14. The function f cuts the strip in half

If one has a function $f : D \longrightarrow R$ and another function $g : R \longrightarrow S$, the *composition* $g \circ f : D \longrightarrow S$ is defined, for any $x \in D$, by

$$g \circ f(x) = g(f(x)) \in S.$$

The function $f : D \longrightarrow D$ defined by $f(x) = x$ is called the *identity function* and is denoted by 1_D.

▷ **Exercise 2.25.** Prove that the composition of two continuous functions is continuous.

(2.18) Definition. *A function $f : D \longrightarrow R$ is invertible if there is a function $g : R \longrightarrow D$ so that*

$$1_R = f \circ g : R \longrightarrow R$$

and

$$1_D = g \circ f : D \longrightarrow D$$

The function g is called the inverse of f, and is usually denoted by f^{-1}.

The set-theoretic inverse f^{-1} of f is a function if and only if f is 1-1 (or a *monomorphism*): $x \neq y$ implies that $f(x) \neq f(y)$. Another way of stating this is: f^{-1} is a function if $f(x) = f(y)$ implies that $x = y$. If $f : A \longrightarrow B$ has an inverse, the inverse will be a function from B to A if f is *onto* or an *epimorphism*: every $y \in B$ is a value for f, so every $y \in B$ satisfies $y = f(x)$ for some $x \in A$. Thus, a function $f : A \longrightarrow B$ is invertible if and only if f is 1-1 and onto.

The example of Figure 2.15 in which f folds the rectangle A along the dotted line is not invertible, since if an inverse f^{-1} did exist it would take the square $f(A)$ and give A back, but to do that the point $\mathbf{x}$ would have to go to both of the points pictured in A. Thus, f^{-1} is not a function.

Fig. 2.15. Folding the rectangle is continuous, but not invertible

Note that the example of Figure 2.14 is 1-1 and onto and thus invertible. The inverse would take the two squares and glue them back together. However, the function is not continuous.

(2.19) Definition. *A function $f : X \longrightarrow Y$ is a homeomorphism if f is continuous and invertible and the inverse function f^{-1} is also continuous. The spaces X and Y are then topologically equivalent.*

An example is the function $f(x) = \tan\left(\frac{\pi x}{2}\right)$ which takes the interval $(-1, 1)$ to $(-\infty, +\infty)$. This function is continuous and has continuous inverse $f^{-1}(x) = \frac{2\tan^{-1}(x)}{\pi}$. Thus, the interval $(-1, 1)$ is topologically equivalent to $(-\infty, \infty) = \mathbb{R}$.

(2.20) Theorem. *Topological equivalence is an equivalence relation.*

Proof. Recall Definition 1.1. Three things must be shown: reflexivity, symmetry, and transitivity. We write $A \sim B$ if there is a continuous function $f : A \longrightarrow B$ with continuous inverse $f^{-1} : B \longrightarrow A$.

(1) Note that $A \sim A$ since the identity function $1_A : A \longrightarrow A$ is continuous and is its own inverse.

(2) If $A \sim B$, then $B \sim A$ since $f^{-1} : B \longrightarrow A$ is continuous and has continuous inverse $f : A \longrightarrow B$.

(3) If $A \sim B$, there is a continuous function $f : A \longrightarrow B$ with continuous inverse $f^{-1} : B \longrightarrow A$, and if $B \sim C$, there is a continuous function $f' : B \longrightarrow C$ with continuous inverse $f'^{-1} : C \longrightarrow B$. Then the compositions $f' \circ f : A \longrightarrow C$ and $f^{-1} \circ f'^{-1} : C \longrightarrow A$ are continuous by Exercise 2.25.

$$(f^{-1} \circ f'^{-1}) \circ (f' \circ f) = f^{-1} \circ (f'^{-1} \circ f') \circ f$$
$$= f^{-1} \circ (1_B) \circ f$$
$$= f^{-1} \circ f = 1_A$$

$$(f' \circ f) \circ (f^{-1} \circ f'^{-1}) = f' \circ (f \circ f^{-1}) \circ f'^{-1}$$
$$= f' \circ (1_B) \circ f'^{-1}$$
$$= f' \circ f'^{-1} = 1_C$$

so $f' \circ f$ and $f^{-1} \circ f'^{-1}$ are inverses. Thus, $f' \circ f$ is a homeomorphism from A to C, so $A \sim C$.

$\square$

Theorem 2.20 implies that one can use topological equivalence like equality, in the context of this course. Topologically equivalent spaces have the same (topological) shape. For example, we have shown that $(-1, 1)$ is topologically equivalent to $\mathbb{R}^1$.

The example above of $f(x) = \tan\left(\frac{\pi x}{2}\right)$ can be modified to give a homeomorphism $f : [0, 1) \longrightarrow [0, \infty)$ by restricting the domain to the half-open interval $[0, 1)$, since $f(0) = 0$. We can then use this homeomorphism to build another homeomorphism $F : D^n(\mathbf{0}, 1) \longrightarrow \mathbb{R}^n$. Any point $\mathbf{x}$ in $D^n(\mathbf{0}, 1)$ can be considered as a vector, and

$$\mathbf{x} = \|\mathbf{x}\| \cdot \frac{\mathbf{x}}{\|\mathbf{x}\|}$$

where $\|\mathbf{x}\|$ is the length of the vector $\mathbf{x}$. Since $\mathbf{x}$ is a vector in $D^n(\mathbf{0}, 1)$, note that $0 \leq \|\mathbf{x}\| < 1$. The quantity $\frac{\mathbf{x}}{\|\mathbf{x}\|}$ is a vector in the same direction as $\mathbf{x}$, but scaled to have length one. Define $F : D^n(\mathbf{0}, 1) \longrightarrow \mathbb{R}^n$ by

$$F(\mathbf{x}) = \begin{cases} f(\|\mathbf{x}\|) \cdot \frac{\mathbf{x}}{\|\mathbf{x}\|}, & \mathbf{x} \neq \mathbf{0} \\ \mathbf{0}, & \mathbf{x} = \mathbf{0} \end{cases}$$

Thus, $F(\mathbf{x})$ is a vector in the same direction as $\mathbf{x}$, but with the length magnified by the function $f(\|\mathbf{x}\|) = \tan\left(\frac{\pi\|\mathbf{x}\|}{2}\right)$ from the example above. The function F can be shown to be continuous and invertible, with continuous inverse. Thus, F takes the disc of radius one homeomorphically to the disc with infinite radius.

Let $\mathbb{S}^2$ denote the sphere or the skin of a ball. Another example of a homeomorphism is defined next.

(2.21) Definition. *Let $\mathbf{x}$ be a point on the sphere $\mathbb{S}^2$. Then $\mathbb{S}^2 - \{\mathbf{x}\}$ is homeomorphic to $\mathbb{R}^2$ by stereographic projection. This is defined by placing the sphere on the plane so that they are tangent and the puncture is at the North Pole. For each $\mathbf{y} \in \mathbb{S}^2$, note that the ray from $\mathbf{x}$ through $\mathbf{y}$ intersects the plane $\mathbb{R}^2$ at a unique point. Define $s(\mathbf{y})$ to be the point where this ray intersects the plane.*

Stereographic projection can be thought of as stretching open the punctured sphere and laying it out flat. (See Fig. 2.16.)

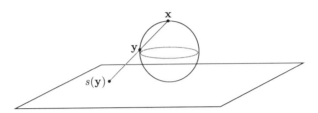

Fig. 2.16. Stereographic projection

▷ **Exercise 2.26.**   Describe what stereographic projection does to

(1) the equator
(2) a longitudinal line from the North Pole to the South Pole
(3) a triangle drawn on the punctured sphere.

Note that we have shown that two spaces, one bounded and one un-bounded, may be topologically equivalent. We wish to identify properties which homeomorphic spaces will share.

(2.22) Definition. *Property P is a topological property if, whenever set A has property P and set B is topologically equivalent to A, then B also has property P.*

Thus, boundedness is not a topological property.

2.4 Compact sets

Although boundedness fails to be a topological property, a related notion does qualify:

(2.23) Definition. *A set A is (sequentially) compact if every infinite sequence of points in A has a limit point in A. That is, if $\{x_i\}_{i=1}^{\infty}$ is a sequence and $x_i \in A$ for each i, then there is a point $x \in A$ so that x is a limit point of $\{x_i\}_{i=1}^{\infty}$.*

There is another type of compactness, which will be briefly described in Chapter 3. Note that Definition 2.23 contains two claims: first, that every sequence has (at least one) limit point, and second, that this limit point will actually lie inside the set.

The real line, $\mathbb{R}$, is not compact, since the sequence $\{1, 2, 3, 4, \dots\} = \{n\}_{n=1}^{\infty}$ consists of points in $\mathbb{R}$, but has no limit point in $\mathbb{R}$. Another example is the interval $(0, 1)$. The number 0 is a limit point for the sequence $\{\frac{1}{2}, \frac{1}{3}, \frac{1}{4}, \frac{1}{5}, \dots\} = \{\frac{1}{n}\}_{n=2}^{\infty} \subseteq (0, 1)$, but $0 \notin (0, 1)$, so the open interval is not compact.

Compactness can be described by terms of properties we have already studied:

(2.24) Heine-Borel Theorem. *A set $A \subseteq \mathbb{R}^n$ is compact if and only if it is closed and bounded.*

Proof. First let us assume that A is a compact set in $\mathbb{R}^n$ and show that A is closed. We must show that if a point $\mathbf{x}$ is chosen with $\mathbf{x} \notin A$, then $\mathbf{x}$ is exterior to A. Since $\mathbf{x} \notin A$, $\mathbf{x}$ cannot be a limit point for any sequence of points in A (otherwise, since A is compact, $\mathbf{x}$ would have to be in A). If every disc around $\mathbf{x}$ contained points of A, then we could construct a sequence as we did in Theorem 2.11 with limit point $\mathbf{x}$. Therefore, there must be some disc $D^n(\mathbf{x}, r)$ with $D^n(\mathbf{x}, r) \cap A = \emptyset$. Thus, $\mathbf{x}$ is exterior to A, so A is closed.

We must also show that if A is compact, then A is bounded. If A is not bounded, then A sprawls out of the disc $D^n(\mathbf{0}, k)$ for every integer k. Thus, for each k there is always at least one point $\mathbf{x}_k$ such that $\mathbf{x}_k \in A$ and $\mathbf{x}_k \notin D^n(\mathbf{0}, k)$ and so $\|\mathbf{x}_k\| > k$. Since $\lim_{k \to \infty} \|\mathbf{x}_k\| \geq \lim_{k \to \infty} k = \infty$, we have a sequence in A with no limit point. This contradicts the assumption that A is compact, so A must be bounded.

The other implication, that if A is closed and bounded, then A is compact, requires two lemmata.

(2.25) Lemma. *The cube $I^n = \{\mathbf{x} = (x_1, x_2, \dots x_n) : -1 \leq x_i \leq 1\} \subseteq \mathbb{R}^n$ is compact.*

Proof. Let $\{\mathbf{x}_k\}_{k=1}^{\infty}$ be any sequence of points in I^n. We must show that $\{\mathbf{x}_k\}$ has a limit point $\mathbf{x}$ and that $\mathbf{x} \in I^n$. Let $I_1^n = I^n$. Divide I_1^n into 2^n equal smaller cubes, as pictured in Figure 2.17.

The infinitely many points of the sequence $\{\mathbf{x}_k\}_{k=1}^{\infty}$ are divided among these 2^n cubes, so at least one of these smaller cubes must contain infinitely many of the $\mathbf{x}_k$'s. Call this cube I_2^n. Repeat the process to obtain a sequence of smaller and smaller nested cubes each containing infinitely many points of the sequence.

For $I^1 = [-1, 1]$, we let $a_1 = -1$, $b_1 = 1$. Divide the interval in half, and choose the half with infinitely many of the $\mathbf{x}_k$'s. If both halves have infinitely many, just choose whichever you like best. Then either $a_2 = -1$ and $b_2 = 0$, or $a_2 = 0$ and $b_2 = 1$, depending on your choice. Take the interval $[a_2, b_2]$ and repeat to get a_k and b_k the endpoints of the kth subinterval selected, Note that $b_k = a_k + \frac{1}{2^{k-2}}$. Thus, $\lim_{k \to \infty} b_k = \lim_{k \to \infty} a_k$, assuming that the limits exist. The a_k's form a bounded monotone sequence and we take

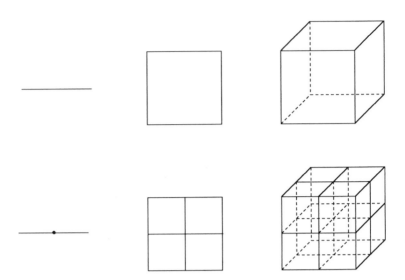

Fig. 2.17. The first subdivision of I^1, I^2, and I^3

as axiomatic, by the completeness of the real numbers, that such sequences must have limits. Let $x = \lim_{k \to \infty} a_k$, and note that $\bigcap_{k=1}^{\infty} I_k$ consists of only one point, x.

For higher dimensions, to show that $\bigcap_{k=1}^{\infty} I_k^n$ is a single point $\mathbf{x}$, repeat the argument for I^1 above for each coordinate. We can, thus, show that $\mathbf{x}$ is a limit point of the sequence $\{\mathbf{x}_k\}$. Since $\mathbf{x} \in I_k^n \subseteq I^n$ for each k, the point $\mathbf{x} \in I^n$ and so I^n is compact.   □

(2.26) Lemma. *If B is a compact set and A is a closed subset of B, then A is compact.*

▷ **Exercise 2.27.**   Prove this lemma. Obviously, since this exercise is placed in the middle of the proof of Theorem 2.24, you are not allowed to use the theorem to prove the lemma.

Proof of Theorem 2.24 continued.   Assume A is closed and bounded. By changing units we may assume $A \subseteq D^n(\mathbf{0}, 1)$ (1 can be 1 inch, 1 foot, 1 mile, whatever). Thus, $A \subseteq D^n(\mathbf{0}, 1) \subseteq I^n$, so A is a closed subset of I^n. The cube I^n is compact by Lemma 2.25, so A is compact by Lemma 2.26.   □

The Heine-Borel Theorem is a fundamental theorem of point-set topology and provides a characterization of compactness in terms previously studied, at least for subsets of $\mathbb{R}^n$.

▷ **Exercise 2.28.**   Show that compactness is a topological property and give examples to show that closedness and boundedness are not.

Compactness is an important property and is the topologists' form of finiteness, in the sense that a compact set does not go on forever. One's first impression is that a bounded set should be "finite," but we have seen that the interval $(-1, 1)$ is topologically equivalent to $(-\infty, \infty)$. One way of thinking about this is to imagine walking along the interval $(-1, 1)$, towards 1, but for some reason (an increase in gravity, or the unnerving fact that one's legs seem to be getting shorter and shorter), the closer one gets to 1, the smaller steps one has to take, so that one never reaches 1. In this sense, $(-1, 1)$ is endless. On the other hand, $[-1, 1]$ is compact and does not go on forever, since it has ends. No topological property should be based solely on distance, as is boundedness, since in topology distance means very little.

2.5 Connected sets

Another fundamental notion is the number of pieces, or *components*, an object has. If one thinks about what "connected" means, an object X is connected, or has only one piece, if all its parts are stuck to each other. Now take the idea "stuck to each other" and look at Figure 2.18.

Fig. 2.18. X is connected, but Y is not

X and Y each have been divided into two parts, where B looks exactly like B', and A looks like A' with the point **x** added on. The point **x** sort of glues A and B together to make X connected, and, being absent from Y, is the gap where Y falls apart into two pieces. Note that **x** is in the set A and is a limit point of both A and B, so that **x** is infinitely close to the set B; in a sense, **x**, and so A, cannot be separated from B. This is how the formal definition of connectedness is derived.

(2.27) Definition. *A set S is connected if whenever S is divided into two non-empty disjoint sets, so that*

$$S = A \cup B, \ A \neq \emptyset, \ B \neq \emptyset, \ A \cap B = \emptyset$$

then either A or B contains a limit point of the other.

Note that in Definition 2.27 all possible divisions of S must be considered. For example, the set Y of Figure 2.18 can be divided in a different way into sets C and D as pictured in Figure 2.19.

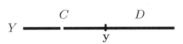

Fig. 2.19. Y is not connected, but this division cannot tell that

Then $Y = C \cup D$, with $\mathbf{y} \in C$, and $C \cap D = \emptyset$, $C \neq \emptyset$, $D \neq \emptyset$, but C does contain a limit point of D. This particular division of Y does not see the gap, but the earlier division of $Y = A' \cup B'$ did.

(2.28) Theorem. *The interval $[0, 1]$ is connected.*

Proof. Assume that the interval $[0, 1]$ is divided into two sets A and B so that $A \neq \emptyset$, $B \neq \emptyset$, $A \cup B = [0, 1]$, and $A \cap B = \emptyset$. Also assume $0 \in A$. Choose some point $b \in B$ (b exists since $B \neq \emptyset$). Let $x_1 = 0$ and $y_1 = b$. Consider the intervals $[0, b/2]$ and $[b/2, b]$. The point $b/2$ is either in A or in B, since $[0, 1] = A \cup B$. If $b/2$ is in A, let $x_2 = b/2$ and $y_2 = b$. If $b/2$ is in B, let $x_2 = 0$ and $y_2 = b/2$. In either case, note that $x_2 \in A$ and $y_2 \in B$ and are the endpoints of the chosen subinterval which has endpoints in both A and B. Take the subinterval $[x_2, y_2]$ and divide it in half again. Of these two sub-subintervals one must have endpoints from both A and B, so call these endpoints x_3 and y_3 where $x_3 \in A$ and $y_3 \in B$. By repeating the bisection infinitely many times, we get two sequences with $\{x_i\}_{i=1}^{\infty}$ a sequence in A and $\{y_i\}_{i=1}^{\infty}$ a sequence in B, and x_i and y_i are endpoints of an interval of decreasing length. Note that

$$\|x_i - y_i\| = \frac{b}{2^{i-1}}$$

Since the interval $[0, 1]$ is compact by Lemma 2.25, these sequences must have limit points. Let x be the limit point of $\{x_i\}_{i=1}^{\infty}$ and y be the limit point of $\{y_i\}_{i=1}^{\infty}$.

$$\|x - y\| = \lim_{n \to \infty} \|x_n - y_n\| = \lim_{n \to \infty} \frac{b}{2^{n-1}} = 0$$

Therefore, $x = y$. The point x is either in A and is a limit point of the sequence $\{y_i\}_{i=1}^{\infty}$ in B and so is a limit point of B, or x is in B and a limit point of A. Thus, $[0, 1]$ is connected.   □

(2.29) Theorem. *If $f : D \longrightarrow R$ is a continuous function from a connected set D onto the set R, then R is connected.*

Proof. If $R = A \cup B$ where $A \neq \emptyset$, $B \neq \emptyset$, $A \cap B = \emptyset$, we must show that either A or B contains a limit point of the other. Let

$$E = \{\mathbf{x} \in D : f(\mathbf{x}) \in A\} = f^{-1}(A)$$

$$F = \{\mathbf{x} \in D : f(\mathbf{x}) \in B\} = f^{-1}(B)$$

Note that since $A \neq \emptyset$, $E \neq \emptyset$ and similarly $B \neq \emptyset$, $F \neq \emptyset$. Also $A \cap B = \emptyset$ implies that $E \cap F = \emptyset$. The domain D is connected, so either E or F contains a limit point of the other. Assume E contains a point $\mathbf{x}$ with $\mathbf{x}$ a limit point of F. Since $\mathbf{x} \in E$, we have $f(\mathbf{x}) \in f(E) = f(f^{-1}(A)) = A$. Since $\mathbf{x}$ is a limit point of the set F and f is continuous, by Theorem 2.16 $f(\mathbf{x})$ is a limit point of the set $f(F) = f(f^{-1}(B)) = B$. Thus, A contains a point $f(\mathbf{x})$ which is a limit point of B. Therefore, R is connected.   □

It follows that connectedness is a topological property.

▷ **Exercise 2.29.**   Prove that X is connected if and only if X cannot be written as a union of two non-empty disjoint sets which are open relative to X.

▷ **Exercise 2.30.**   Prove that X is connected if and only if X cannot be written as a union of two non-empty disjoint sets which are closed relative to X.

▷ **Exercise 2.31.**   Give examples of sets A and B in $\mathbb{R}^2$ which satisfy:

(1) A and B are connected, but $A \cap B$ is not connected
(2) A and B are connected, but $A - B$ is not connected
(3) A and B are each not connected, but $A \cup B$ is connected.

2.6 Applications

In this section, we use the properties we have learned, continuity and connectedness, to derive some rather amusing and practical applications.

(2.30) Theorem. *If* $f : [0, 1] \longrightarrow [0, 1]$ *is a continuous function, then there is a point* $t \in [0, 1]$ *with* $f(t) = t$.

Proof. Assume that $f(x)$ is never equal to x. Then consider the function $g : [0, 1] \to \{\pm 1\}$ defined by

$$g(x) = \frac{f(x) - x}{|f(x) - x|}$$

Note that the quantity $\frac{a}{|a|}$ is equal to $+1$ if a is positive, and -1 if a is negative. Since $f(x) \neq x$, we have not divided by zero in the definition of g, and the continuity of g follows from the continuity of f. Since we have assumed that $f(0) \neq 0$, it must be that $f(0) > 0$ and so $g(0) = +1$. Similarly, $f(1) \neq 1$, so $f(1) < 1$ and $g(1) = -1$. Thus, g is a continuous function taking the interval $[0, 1]$ onto the set $\{-1, 1\}$. By Theorem 2.28, the interval $[0, 1]$ is connected. Theorem 2.29 states that a continuous function must take a connected set to another connected set. However, g takes $[0, 1]$ to $\{-1, 1\}$, which is not connected. There is a contradiction unless there is a point $t \in [0, 1]$ with $f(t) = t$.   □

The theorem above is the one-dimensional version of a famous theorem, the Brouwer Fixed Point Theorem. This will be proven for higher dimensions in Chapter 10. Points such as t above with $f(t) = t$ are called *fixed points*, since f does not move t.

▷ **Exercise 2.32.** Let $f : \mathbb{R}^n \longrightarrow \mathbb{R}^n$ be a continuous function. Show that the *fixed point set* for f, $\mathfrak{F}(f) = \{t \in \mathbb{R}^n : f(t) = t\}$, is closed.

(2.31) Definition. *A space X has the fixed point property if every continuous function $f : X \longrightarrow X$ has a fixed point.*

Theorem 2.30 above shows that the interval $[0, 1]$ has the fixed point property.

(2.32) Theorem. *The fixed point property is a topological property.*

▷ **Exercise 2.33.** Prove Theorem 2.32. [Hint: let X be a space with the fixed point property, and let Y be another space with a homeomorphism $f : Y \longrightarrow X$. Prove that if $g : Y \longrightarrow Y$ is any continuous function, then there is a point $y \in Y$ such that $g(y) = y$, by using the invertible function f to transfer the question to X.]

Let $\mathbb{S}^1$ denote the unit circle, i.e., $\mathbb{S}^1 = \{(x, y) \in \mathbb{R}^2 : x^2 + y^2 = 1\}$.

▷ **Exercise 2.34.** Give examples of functions which show that $\mathbb{R}$ and $\mathbb{S}^1$ do not have the fixed point property.

In higher dimensions, the n-dimensional sphere is defined as

$$\mathbb{S}^n = \{\mathbf{x} = (x_1, x_2, \dots, x_{n+1}) \in \mathbb{R}^{n+1} : \|\mathbf{x}\| = 1\}.$$

(2.33) Definition. *Consider the n-sphere $\mathbb{S}^n$ as a subset of $\mathbb{R}^{n+1}$. The points $\mathbf{x} = (x_1, x_2, \dots, x_{n+1})$ and $-\mathbf{x} = (-x_1, -x_2, \dots, -x_{n+1})$ in $\mathbb{S}^n$ are called antipodal.*

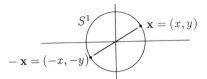

Fig. 2.20. Antipodal points in $\mathbb{S}^1$

Thus, antipodal points are diametrically opposite each other (see Fig. 2.20). The next theorem is the one-dimensional version of another well-known theorem, the Borsuk-Ulam Theorem, about functions on the circle.

(2.34) Theorem. *If f is a continuous function from the circle $\mathbb{S}^1$ to a subset of $\mathbb{R}$, then there is a pair of antipodal points $\mathbf{x}$ and $-\mathbf{x}$ on $\mathbb{S}^1$ with $f(\mathbf{x}) = f(-\mathbf{x})$. In other words, there is a pair of antipodal points where f has the same value.*

Proof. Assume $f(\mathbf{x}) \neq f(-\mathbf{x})$ for all $\mathbf{x} \in \mathbb{S}^1$. Define a new function g by

$$g(\mathbf{x}) = \frac{(f(\mathbf{x}) - f(-\mathbf{x}))}{|f(\mathbf{x}) - f(-\mathbf{x})|}$$

Then g is continuous, since f is continuous and we have not divided by zero. Note that

$$g(-\mathbf{x}) = \frac{(f(-\mathbf{x}) - f(\mathbf{x}))}{|f(-\mathbf{x}) - f(\mathbf{x})|} = -g(\mathbf{x})$$

Thus, if $g(\mathbf{x}) = +1$, then $g(-\mathbf{x}) = -1$, and vice versa. We have constructed a continuous function g from the connected space $\mathbb{S}^1$ onto the set $\{-1, 1\}$ which is not connected. This gives a contradiction unless there is, after all, a point $\mathbf{x}$ where $f(\mathbf{x}) = f(-\mathbf{x})$. $\quad\Box$

(2.35) Corollary. *At any given moment, there are two points on the equator exactly opposite each other with the same temperature.*

Proof. The equator is a circle $\mathbb{S}^1$. Let f be the temperature function and use Theorem 2.34. $\quad\Box$

The following theorem is familiarly known as the pancake theorem, since it describes how to divide pancakes in half. Think of A as consisting of several (infinitely thin) pancakes and B a pool of maple syrup. The theorem then implies, but does not tell one how to locate, that there is a single knife slice dividing everything into two equal breakfasts.

(2.36) Theorem. *Let A and B be bounded subsets of $\mathbb{R}^2$. They may overlap and need not be connected. There is a line which divides each region in half.*

Proof. Let S be a circle centered at the origin so that A and B are contained within S. We can, by changing units of measurement, assume that S has diameter 1. For any point $\mathbf{x} \in S$, let $D_{\mathbf{x}}$ be the diameter through $\mathbf{x}$, and let L_t be the line perpendicular to $D_{\mathbf{x}}$ at distance t from $\mathbf{x}$, where $0 \le t \le 1$. See Figure 2.21.

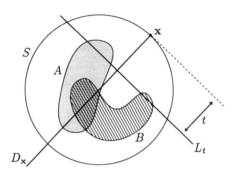

Fig. 2.21. $D_{\mathbf{x}}$ and L_t

Define two functions:

$$f_1(t) = \text{area of the portion of } A \text{ on the } \mathbf{x}\text{-side of } L_t$$
$$f_2(t) = (\text{total area of } A) - f_1(t)$$
$$= \text{area of the portion of } A \text{ on the far side of } L_t.$$

Note that $f_1(0) = 0$ and $f_2(1) = 0$, and that f_1 and f_2 are continuous functions from $[0,1]$ to a subset of $\mathbb{R}$. Define a new function $f : [0,1] \longrightarrow \mathbb{R}$ by

$$f(s) = f_2(s) - f_1(s)$$

Since f_1 and f_2 are continuous, so is f, and

$$f(0) = f_2(0) - f_1(0)$$
$$= f_2(0) - 0$$
$$= \text{total area of } A$$
$$f(1) = f_2(1) - f_1(1)$$
$$= 0 - f_1(1)$$
$$= -(\text{total area of } A)$$

If $f(s)$ is never zero, then consider $\frac{f(s)}{|f(s)|} = \pm 1$. This would be a continuous function of $[0,1]$ to $\{-1,1\}$, which is impossible. Thus, there must be a point $t_{\mathbf{x}}$ with $f(t_{\mathbf{x}}) = 0$, so $f_1(t_{\mathbf{x}}) = f_2(t_{\mathbf{x}})$. The line $L_{t_{\mathbf{x}}}$ drawn $t_{\mathbf{x}}$ units from $\mathbf{x}$ on the line $D_{\mathbf{x}}$ bisects A.

The line $L_{t_{\mathbf{x}}}$ also divides the region B into two (probably not equal) parts. Do this for each point $\mathbf{x} \in S$. Define two more functions:

$$g_1(\mathbf{x}) = \text{area of the portion of } B \text{ on the } \mathbf{x}\text{-side of } L_{t_{\mathbf{x}}}$$
$$g_2(\mathbf{x}) = \text{area of the portion of } B \text{ on the far side of } L_{t_{\mathbf{x}}}$$

Let $g(\mathbf{x}) = g_1(\mathbf{x}) - g_2(\mathbf{x})$. Then g is a continuous function from the circle S to $\mathbb{R}$. Note that for any point $\mathbf{x} \in S$, the line $D_{\mathbf{x}}$ is the same line as $D_{-\mathbf{x}}$. For any $\mathbf{x} \in S$,

$$
\begin{aligned}
g_1(\mathbf{x}) &= \text{area of the portion of } B \text{ on the } \mathbf{x}\text{-side of } L_{t_{\mathbf{x}}} \\
&= \text{area of the portion of } B \text{ on the far side from } -\mathbf{x} \\
&= g_2(-\mathbf{x}) \\
g_2(\mathbf{x}) &= \text{area of the portion of } B \text{ on the far side of } L_{t_{\mathbf{x}}} \text{ from } \mathbf{x} \\
&= \text{area of the portion of } B \text{ on the } (-\mathbf{x})\text{-side of } L_{t_{\mathbf{x}}} \\
&= g_1(-\mathbf{x})
\end{aligned}
$$

Thus, for all $\mathbf{x} \in S$,

$$g(-\mathbf{x}) = g_1(-\mathbf{x}) - g_2(-\mathbf{x}) = g_2(\mathbf{x}) - g_1(\mathbf{x}) = -g(\mathbf{x})$$

By Theorem 2.34, there is a point $\mathbf{y} \in S$ with $g(\mathbf{y}) = g(-\mathbf{y})$. For the point $\mathbf{y}$, we have $g(\mathbf{y}) = g(-\mathbf{y}) = -g(\mathbf{y})$, so $g(\mathbf{y})$ must be 0. Therefore, $g(\mathbf{y}) = 0 = g_1(\mathbf{y}) - g_2(\mathbf{y})$, so $g_1(\mathbf{y}) = g_2(\mathbf{y})$. At the point $\mathbf{y} \in S$, B is also bisected by the line $L_{t_{\mathbf{y}}}$. $\qquad\square$

Chapter 3

Point-set topology

"I preach mathematics; who will occupy himself with the study of mathematics will find it the best remedy against the lusts of the flesh."

Thomas Mann

3.1 Open sets and neighborhoods

Remember that all of the examples and theorems in the previous chapter dealt with sets in $\mathbb{R}^n$ and inherited a lot of structure from the standard euclidean structure of $\mathbb{R}^n$. In particular, the definition of our neighborhoods $D^n(\mathbf{x}, r) = \{\mathbf{y} \in \mathbb{R}^n : \|\mathbf{y} - \mathbf{x}\| < r\}$ makes use of the euclidean distance. It is possible to study point-set topology on a much more abstract level, by using different neighborhoods. Notice that all the definitions in Chapter 2 were based on the concept of a neighborhood of a point or on the concept of an open set. The definitions of interior, limit point, closed set, connectedness, and continuity all can be rewritten to depend only on the ideas of neighborhoods and open sets. We defined a neighborhood to be an open disc around a point, but this choice was really arbitrary.

The standard topology on $\mathbb{R}^n$ is an example of the following construction:

(3.1) Definition. *Let X be a set with a function d defined for $x, y \in X$, so that $d(x, y) \in [0, \infty)$ and d satisfies:*

    (1) $d(x, y) = 0$ *if and only if $x = y$*
    (2) $d(x, y) = d(y, x)$ *for each $x, y \in X$*
    (3) $d(x, z) \le d(x, y) + d(y, z)$ *for each $x, y, z \in X$.*

Then X is a metric space with metric d.

(3.2) Definition. *Let X be a metric space. The metric topology on X is defined by using sets N as neighborhoods of a point $x \in X$ where*

$$N = D_X(x, r) = \{y \in X : d(x, y) < r\}$$

and r is a real number greater than 0.

Property (3) of Definition 3.1 is called the triangle inequality. These properties are obvious properties of the euclidean distance function. Thus, the standard topology on $\mathbb{R}^n$ is the metric topology for the standard euclidean metric. Any distance function defines a collection of neighborhoods, which, in turn, defines open and closed sets, continuous functions, etc.

However, it is not always easy or even possible to find a metric, and this approach is philosophically repugnant, since it violates one of the fundamental notions of topology: the neighborhoods are defined in terms of a measure of distance. In topology, distances should not matter, since we consider everything to be elastic. We can remove this objection by redefining everything in terms of more general open sets and neighborhoods.

(3.3) Definition. *A topological space is a set X with a collection $\mathcal{B}$ of subsets $N \subseteq X$, called neighborhoods, such that*

(1) *every point is in some neighborhood, i.e.,*

$$\forall x \in X, \exists N \in \mathcal{B} \text{ such that } x \in N$$

(2) *the intersection of any two neighborhoods of a point contains a neighborhood of the point, i.e.,*

$$\forall N_1, N_2 \in \mathcal{B} \text{ with } x \in N_1 \cap N_2, \exists N_3 \in \mathcal{B} \text{ such that } x \in N_3 \subseteq N_1 \cap N_2$$

The set, $\mathcal{B}$, of all neighborhoods is called a basis for the topology on X.

(3.4) Definition. *Let X be a topological space with a basis $\mathcal{B}$. A subset $O \subseteq X$ is an open set if for each $x \in O$, there is a neighborhood $N \in \mathcal{B}$ such that $x \in N$ and $N \subseteq O$. The set $\mathcal{T}$ of all open sets is a topology on the set X.*

Note that this is essentially Definition 2.3 of the previous chapter. It is clear that any neighborhood is itself an open set by this definition. It is possible to define everything using only open sets, but the smaller collection of neighborhoods is usually easier to work with. It is also possible to take the collection of all open sets as a basis for the topology, so that a neighborhood is by definition any open set.

We first verify that this abstract definition of an open set satisfies the properties that one expects an open set to have.

(3.5) Theorem. *Let X be a topological space with topology $\mathcal{T}$ and basis $\mathcal{B}$.*

  (1) *X and $\emptyset$ are elements of $\mathcal{T}$.*
  (2) *The union of any collection of elements in $\mathcal{T}$ is in $\mathcal{T}$.*
  (3) *The intersection of any finite collection of elements in $\mathcal{T}$ is in $\mathcal{T}$.*

Proof.

  (1) Since each neighborhood $N \subseteq X$ and Definition 3.3(1) states that for each $x \in X$, $x \in N$ for some $N \in \mathcal{B}$, it follows that each $x \in X$ satisfies $x \in N \subseteq X$ for some $N \in \mathcal{B}$. Thus, X is open. The null set, $\emptyset$, is considered open since there are no $x \in \emptyset$, so Definition 3.4 is vacuously true.

  (2) If $U = \bigcup_\alpha O_\alpha$ for some collection $\{O_\alpha\} \subseteq \mathcal{T}$, then for any $x \in U$, $x \in O_\alpha$ for one of the open sets O_α, so there is a neighborhood N with $x \in N \subseteq O_\alpha$. Then $x \in N \subseteq O_\alpha \subseteq \bigcup_\alpha O_\alpha = U$, so U is open and is in $\mathcal{T}$.

  (3) If $U = \bigcap_{i=1}^{n} O_i$ for some finite collection $\{O_i\}_{i=1}^n \subseteq \mathcal{T}$, let $x \in U$. Then $x \in O_i$ for each i, and since each O_i is open, there is a neighborhood N_i with $x \in N_i \subseteq O_i$. Note that by Definition 3.3(2), since $x \in N_1 \cap N_2$, there is a neighborhood N' with $x \in N' \subseteq N_1 \cap N_2$. Repeat for $x \in N' \cap N_3$, etc. The process is finite, so one ends with x in a neighborhood N and

$$x \in N \subseteq \bigcap_{i=1}^{n} N_i \subseteq \bigcap_{i=1}^{n} O_i = U$$

so U is open.

$\square$

Note that we proved in Chapter 2 that the open sets in $\mathbb{R}^n$ have these properties (see Exercises 2.13, 2.14, 2.19, 2.20).

(3.6) Theorem. *Let X be a topological space with basis $\mathcal{B}$ and topology $\mathcal{T}$. A set $O \subseteq X$ is open (so $O \in \mathcal{T}$) if and only if O can be written as a union of elements of $\mathcal{B}$.*

▷ **Exercise 3.1.** Prove Theorem 3.6.

▷ **Exercise 3.2.** Let X be a metric space. Show that the collection of neighborhoods defined in Definition 3.2 forms a basis for the metric topology. That is, show that these neighborhoods satisfy the conditions of Definition 3.3.

Let X be the set of two elements $\{x, y\}$. One can figure out all possible topologies on X, represented schematically in Figure 3.1. Remember that

$\emptyset$ must always be considered as an open set. The first picture indicates the topology $\mathfrak{T} = \{X, \emptyset\}$, etc.

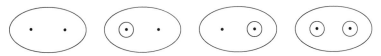

Fig. 3.1. Schematic representations of all the topologies on a set with two elements

▷ **Exercise 3.3.**   Let $X = \{x, y, z\}$, the set with three elements. Figure out as many topologies for X as you can and represent them both schematically and by listing their open sets.

It is convenient to shift the point of view from the relative position of individual points adopted in Chapter 2 (where one focused on whether a point was interior or exterior to a given set, for example) to the properties of sets of points. The properties studied in this chapter are largely generalizations of those of Chapter 2, but we do not always choose to adapt the definition used in the earlier chapter, but often choose a statement equivalent to it. To define the closed sets, we choose the analog of Exercise 2.11 rather than Definition 2.4.

(3.7) Definition. *If C is a subset of a topological space X with topology $\mathfrak{T}$, then C is closed if $X - C$ is open.*

(3.8) Theorem. *Let X be a topological space.*
  (1) X and $\emptyset$ are closed sets.
  (2) The intersection of any collection of closed sets in X is closed.
  (3) The union of any finite collection of closed sets in X is closed.

▷ **Exercise 3.4.**   Prove Theorem 3.8.

Of course, changing the system of neighborhoods used on a space may change which sets are considered to be open, and thus change all the properties that we have studied such as connectedness, compactness, and continuity, since these are defined in terms of open sets. Consider $\mathbb{R}^2$ with a new system of neighborhoods defined by letting a neighborhood of (x, y) be the open rectangle centered at (x, y) with width $2a$ for some a and height $2b$ for some b, i.e., let

$$N = \{(x', y') : x - a \le x' \le x + a, \ y - b \le y' \le y + b\}$$

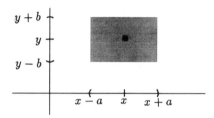

$y + b$

y

$y - b$

$\overset{(}{x-a} \quad \overset{|}{x} \quad \overset{)}{x+a}$

Fig. 3.2. A neighborhood of (x, y) in $\mathbb{R}^2$ with the open rectangle (or product) topology

We need to decide how to compare this topology with the standard topology of $\mathbb{R}^2$ studied in Chapter 2. The topologies will, of course, be the same if the set of all open sets turns out to be the same.

(3.9) Definition. *Let X be a topological space with two bases $\mathcal{B}$ and $\mathcal{B}'$. Let $\mathcal{T}$ and $\mathcal{T}'$ be the collections of all open sets corresponding to these bases. Then $\mathcal{B}$ is equivalent to $\mathcal{B}'$ if $\mathcal{T} = \mathcal{T}'$.*

(3.10) Theorem. *Let X be a topological space with two bases $\mathcal{B}$ and $\mathcal{B}'$. Then $\mathcal{B}$ is equivalent to $\mathcal{B}'$ if and only if*

  (1) *for each $B \in \mathcal{B}$ and each $x \in B$, there is a $B' \in \mathcal{B}'$ such that $x \in B' \subseteq B$;*
  (2) *for each $B' \in \mathcal{B}'$ and each $x \in B'$, there is a $B \in \mathcal{B}$ such that $x \in B \subseteq B'$.*

▷ **Exercise 3.5.**   Prove Theorem 3.10.

▷ **Exercise 3.6.**   Show that the two bases for $\mathbb{R}^2$, the standard euclidean topology of Chapter 2 and that illustrated in Figure 3.2, are equivalent.

Equivalent bases give the same collection of open sets, and so the same closed sets, connected sets, and continuous functions. We can use these bases interchangeably.

A different structure on $\mathbb{R}^1$ is defined by using the *half-open topology.* Let an interval of the form $[a, b)$ be considered as a neighborhood of any $x \in [a, b)$. This is not the same topology as the one discussed in Chapter 2, where neighborhoods of x in $\mathbb{R}^1$ were defined to be intervals of the form $(x - \epsilon, x + \epsilon)$ for some $\epsilon > 0$. To see that these are not equivalent, consider the interval $[0, 1)$. This is considered to be a neighborhood of the point

0 and, thus, open in the half-open topology, but no interval of the form $(-\epsilon, +\epsilon)$ is contained in $[0, 1)$.

Two other examples follow, to show you how far-fetched things can get. The *discrete topology* is defined for X any set. Neighborhoods are defined to be any non-empty set. In particular, $\{x\}$ is a neighborhood of the point x. It then appears, by Theorem 3.6, that any set is open, and it follows from Definition 3.7 that all sets are closed. Another extreme example is the *indiscrete topology*, defined for any set X. The only allowable neighborhood is $N = X$. Then X and $\emptyset$ are the only open sets. These are also the only closed sets.

If X is a topological space with a subset A, then A inherits a topology from X, called the *subspace topology*, defined as the relative topology was defined in Chapter 2.2:

(3.11) Definition. *Let X be a topological space with topology $\mathfrak{T}$ and let $A \subseteq X$. A neighborhood of a point $x \in A$ relative to A is of the form $N \cap A$ where N is a neighborhood of x in X. The topology $\mathfrak{T}_A$ generated by this basis is called the subspace topology on A induced by the topology $\mathfrak{T}$ on X.*

(3.12) Theorem. *Let A be a subset of a topological space X. The open sets relative to A are precisely the open sets of X intersected with A; i.e., B is open in A if and only if $B = A \cap O$ for some set O which is open in X. Furthermore, B is closed in A if and only if $B = A \cap C$ for some set C which is closed in X.*

Proof. Let $B = A \cap O$ for O open in X. If $x \in B$, then $x \in O$, so there is a neighborhood N in X such that $x \in N \subseteq O$. Thus, $x \in N \cap A \subseteq O \cap A = B$. The set $N \cap A$ is a neighborhood for the relative topology on A, so B is open in A.

If B is open in A, then for every $x \in B$, there is a neighborhood N_x so that the relative neighborhood $N_x \cap A$ satisfies $x \in N_x \cap A \subseteq B$. This is true for each $x \in B$ since B is open in A. Thus,

$$B = \bigcup_{x \in B} (N_x \cap A) = \left(\bigcup_{x \in B} N_x \right) \cap A = O \cap A$$

where O is defined by $O = \cup_{x \in B} N_x$, and so is open in X by Theorem 3.5(2). □

▷ **Exercise 3.7.** Prove the assertion about closed sets in Theorem 3.12.

3.2 Continuity, connectedness, and compactness

The definition of a continuous function between topological spaces is easily adapted from Definition 2.15 of the previous chapter:

(3.13) Definition. *Let X and Y be topological spaces. Then a function $f : X \longrightarrow Y$ is continuous if whenever a set A is open in Y, $f^{-1}(A)$ is an open set in X. If we let $\mathcal{T}_X$ denote the topology on X, and $\mathcal{T}_Y$ the topology for Y, this can be restated as: f is continuous if for every $A \in \mathcal{T}_Y$, $f^{-1}(A) \in \mathcal{T}_X$.*

To see how the topology on X and Y affects this definition, let us consider the jump function $f : \mathbb{R} \longrightarrow \{0,1\}$ defined by

$$f(x) = \begin{cases} 0 & \text{if } x < 0 \\ 1 & \text{if } x \geq 0 \end{cases}$$

Let $\{0,1\}$ have the discrete topology, so $\{0\}$ and $\{1\}$ are considered to be open sets. Let us consider $\mathbb{R}$ with two different topologies. First, let $\mathbb{R}$ have the standard topology and note that $f^{-1}\{0\} = (-\infty, 0)$, which is open in the standard topology, but $f^{-1}\{1\} = [0, +\infty)$ which is not open. The jump function is, thus, not continuous with respect to the standard topology. Next consider the same function but change the topology on $\mathbb{R}$ to the half-open topology, and notice that both $(-\infty, 0)$ and $[0, +\infty)$ are open, so that the jump function is continuous with respect to the half-open topology.

▷ **Exercise 3.8.** Let both X and Y have the discrete topology. Show that any function $f : X \longrightarrow Y$ is continuous.

▷ **Exercise 3.9.** Let X and Y be topological spaces. Prove that a function $f : X \longrightarrow Y$ is continuous if and only if for each closed $C \subseteq Y$, $f^{-1}(C)$ is closed in X.

▷ **Exercise 3.10.** Let X be a topological space with closed subsets A and B such that $X = A \cup B$. Let $f : A \longrightarrow Y$ and $g : B \longrightarrow Y$ be continuous functions such that for each $x \in A \cap B$, $f(x) = g(x)$. Define a new function $f \cup g : X \longrightarrow Y$ by

$$f \cup g(x) = \begin{cases} f(x) & \text{for } x \in A \\ g(x) & \text{for } x \in B \end{cases}$$

(1) Prove that $f \cup g$ is continuous.
(2) Give an example to show that the condition that A and B must be closed is necessary.

Connectedness is also easily defined in terms of open sets, by adapting Exercise 2.29:

(3.14) Definition. *A topological space X is connected if X cannot be written as a union of two non-empty disjoint open sets.*

The interval $[0, 1)$ is connected in the standard topology (by slightly modifying the proof of Theorem 2.28), but is not connected in the half-open topology, since $[0, 1) = [0, \frac{1}{2}) \cup [\frac{1}{2}, 1)$ and both $[0, \frac{1}{2})$ and $[\frac{1}{2}, 1)$ are neighborhoods and, thus, open sets.

▷ **Exercise 3.11.** Show that if X is a non-empty topological space with the discrete topology, then the only connected sets are the sets of one element of the form $\{x\}$ for $x \in X$.

▷ **Exercise 3.12.** Show that if X is a non-empty topological space with the indiscrete topology, then any subset of X is connected.

▷ **Exercise 3.13.** Show that the "flea and comb" space defined as $X = \{(0, 1)\} \cup \{(x, 0) : 0 < x \leq 1\} \cup \{(\frac{1}{n}, y) : n \subseteq \mathbb{N}, 0 \leq y \leq 1\}$ is a connected subset of $\mathbb{R}^2$ with the standard topology.

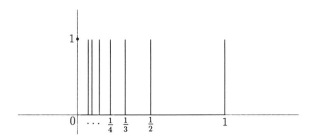

Fig. 3.3. The flea and comb space

(3.15) Theorem. *Let X and Y be topological spaces and $f : X \longrightarrow Y$ a continuous function onto Y. If X is connected, then Y is connected.*

▷ **Exercise 3.14.** Prove Theorem 3.15.

Note that Theorem 3.15 is the analog of Theorem 2.29 and implies that connectedness is a topological property.

In Definition 2.23 we defined a version of compactness called sequential compactness. This was not defined in terms of open sets only, and so we use a different definition:

(3.16) Definition. *Let A be a subset of a topological space X. An open cover of A is a collection $\mathcal{O}$ of open subsets of X so that A lies in the union of the elements of $\mathcal{O}$, i.e.,*

$$A \subseteq \bigcup_{O \in \mathcal{O}} O$$

A subcover of $\mathcal{O}$ is a subcollection $\mathcal{O}' \subseteq \mathcal{O}$ so that A lies in the union of the elements of $\mathcal{O}'$. A finite cover (or subcover) is a cover $\mathcal{O}$ consisting of finitely many sets.

(3.17) Definition. *A topological space X is compact if every open cover of X has a finite subcover.*

▷ **Exercise 3.15.**   Show that if X is a topological space consisting of a finite number of points, with any topology, then X is compact.

The space $\mathbb{R}^2$ is not compact with the standard topology, since the collection $\mathcal{O} = \{D^2(0, n) : n \in \mathbb{Z}\}$ is a cover of $\mathbb{R}^2$ by the open discs of radius n, but no finite subcover will cover all of $\mathbb{R}^2$. Note that since $\mathbb{R}^2$ is itself open, it is, of course, a finite open cover for itself. A space is compact not if it has a finite open cover but if *every* open cover has a finite subcover.

Another example, using the standard topology, is $(0, 1)$ as a subset of $\mathbb{R}$. An open cover of $(0, 1)$ is $\mathcal{O} = \{(\frac{1}{n}, 1) : n \in \mathbb{Z}\}$. Again, no finite subcover will cover all of the interval $(0, 1)$, so the open interval is not compact.

▷ **Exercise 3.16.**   Show that any space X with the indiscrete topology is compact.

The Heine-Borel Theorem 2.24, which stated that a set $A \subseteq \mathbb{R}^n$ is sequentially compact if and only if it is closed and bounded, is true with this new definition of compactness and the standard topology on $\mathbb{R}^n$ but is not true for all topological spaces. As an example of a space which is closed and bounded but not compact, define $\mathbb{R}^\infty$ to be the set of all infinite "points" $\mathbf{x} = (x_1, x_2, x_3, \dots)$ where $x_i \in \mathbb{R}$ and only a finite number of the x_i's are non-zero. A metric on this space is defined for points $\mathbf{x} = (x_1, x_2, x_3, \dots)$ and $\mathbf{y} = (y_1, y_2, y_3, \dots)$ by

$$d(\mathbf{x}, \mathbf{y}) = \sqrt{\sum_{i=1}^{\infty} (x_i - y_i)^2}$$

▷ **Exercise 3.17.**   Prove that the formula above defines a metric on $\mathbb{R}^\infty$.

Give $\mathbb{R}^\infty$ the topology induced by this metric as in Definition 3.2. Define $\mathbf{x}_i = (0, 0, \dots, 0, 1, 0, \dots, 0, \dots) \in \mathbb{R}^\infty$, where $\mathbf{x}_i$ has a 1 in the ith coordinate and 0 everywhere else. Consider the subset $A = \{\mathbf{x}_1, \mathbf{x}_2, \mathbf{x}_3, \dots\}$ of

$\mathbb{R}^\infty$. Note that for any two points $\mathbf{x}_i$ and $\mathbf{x}_j$ in A, the distance between them is

$$d(\mathbf{x}_i, \mathbf{x}_j) = \sqrt{0 + \cdots + 0 + (1-0)^2 + 0 + \cdots + 0 + (0-1)^2 + 0 + \cdots} = \sqrt{2}$$

▷ **Exercise 3.18.** Prove that the set A defined above is bounded.

▷ **Exercise 3.19.** Prove that $\mathbb{R}^\infty - A$ is open, and, thus, A is closed.

▷ **Exercise 3.20.** Prove that A is not compact.

However, compare Theorem 3.18 below with Lemma 2.26:

(3.18) Theorem. *If X is a compact topological space and A is a closed subset of X, then A is compact.*

Proof. Let $\mathcal{O}$ be an open cover of the closed subset A of the topological space X. Then each $O \in \mathcal{O}$ is an open set in X. Since A is closed, $X - A$ is open. Define an open cover $\mathcal{O}^*$ of X by

$$\mathcal{O}^* = \mathcal{O} \cup (X - A)$$

Since X is compact, $\mathcal{O}^*$ has a finite subcover $\mathcal{O}^{*\prime} = \{O_1, O_2, \ldots, O_n\}$, where one of the sets O_i may be $X - A$. If $\mathcal{O}^{*\prime}$ includes $X - A$, let $\mathcal{O}' = \mathcal{O}^{*\prime} - \{X - A\}$, so that $\mathcal{O}' \subseteq \mathcal{O}$. If $\mathcal{O}^{*\prime}$ does not include $X - A$, let $\mathcal{O}' = \mathcal{O}^{*\prime}$, so again $\mathcal{O}' \subseteq \mathcal{O}$. In either case, $\mathcal{O}'$ is a finite subcover of $\mathcal{O}$ and will cover A, since omitting $X - A$ will not leave out any points of A.   □

Also, compactness remains a topological property:

(3.19) Theorem. *Let X be a compact topological space and $f : X \longrightarrow Y$ a continuous function from X onto a topological space Y. Then Y is compact.*

▷ **Exercise 3.21.** Prove Theorem 3.19.

With this new definition, compactness by Definition 3.17 implies the sequential compactness of Definition 2.23, but for certain rather strange spaces the converse is not true. The example above of $\mathbb{R}^\infty$ as a closed and bounded space which is not compact is not sequentially compact either.

3.3 Separation axioms

The *separation axioms* are a way of cataloging how weird a topology is.

(3.20) Definition. *A topological space X is a T_0 space if for every pair of distinct points x, $y \in X$, there is an open set containing one of the points but not the other.*

(3.21) Definition. *A topological space X is a T_1 space if for every pair of distinct points x, $y \in X$, there are open sets U and V so that $x \in U$ but $y \notin U$, and $y \in V$ but $x \notin V$.*

(3.22) Definition. *A topological space X is a T_2 or Hausdorff space if for every pair of distinct points x, $y \in X$, there are disjoint open sets U and V so that $x \in U$ and $y \in V$.*

Here is an example of a T_0 space which is not T_1: Let X consist of two points x and y, and define a topology by letting the open sets be

$$\mathcal{T} = \{\emptyset, \{x\}, X\}$$

Note that this topology satisfies the conditions of Theorem 3.5. The open set $U = \{x\}$ contains x but not y, so this space is T_0. The only open set containing y is $V = X$, so this space is not T_1.

An example of a space that is T_1 but not T_2 can be constructed from the set of natural numbers $\mathbb{N} = \{1, 2, 3, 4, \dots\}$ where a subset U of $\mathbb{N}$ is considered open if either $\mathbb{N} - U$ is finite or $U = \emptyset$. This is called the *finite complement topology*. If x and y are elements of $\mathbb{N}$ and $x \neq y$, then $U = \mathbb{N} - \{y\}$ is an open set containing x but not y, and $V = \mathbb{N} - \{x\}$ is an open set containing y but not x. Thus, this example is T_1. Since open sets must have finite complements, any two open sets must have an infinite number of common elements. One cannot construct disjoint open sets around x and y, and so this space is not Hausdorff.

▷ **Exercise 3.22.**   Show that $\mathbb{N}$ is connected in the finite complement topology.

(3.23) Definition. *Let X be a topological space such that for each $x \in X$, the set $\{x\}$ is closed. Then X is a regular space if for every closed set $C \subseteq X$ and every point $x \notin C$, there are disjoint open sets U and V so that $x \in U$ and $C \subseteq V$.*

(3.24) Definition. *Let X be a topological space such that for each $x \in X$, the set $\{x\}$ is closed. Then X is a normal space if for every pair of disjoint closed sets C and $D \subseteq X$, there are disjoint open sets U and V so that $C \subseteq U$ and $D \subseteq V$.*

Euclidean space $\mathbb{R}^n$ is normal.

▷ **Exercise 3.23.** Prove that

$$\text{Normal} \implies \text{regular} \implies T_2 \implies T_1 \implies T_0$$

▷ **Exercise 3.24.** Determine which separation axioms your examples of Exercise 3.3 satisfy.

▷ **Exercise 3.25.** Prove that if X is a Hausdorff space, Y is a compact subset of X, and $x \in X - Y$, then there are disjoint open sets U and V in X such that $x \in U$ and $Y \subseteq V$.

Exercise 3.25 is used in the proof of the following theorem. Compare this result with Theorem 3.18.

(3.25) Theorem. *If X is a Hausdorff space and Y is a compact subset of X, then Y is closed.*

▷ **Exercise 3.26.** Prove Theorem 3.25.

(3.26) Theorem. *If X is a compact topological space and Z is a Hausdorff space, with a continuous one-to-one function $f : X \longrightarrow Z$ onto Z, then f is a homeomorphism.*

▷ **Exercise 3.27.** Prove Theorem 3.26, using Theorem 3.25 and Exercise 3.9.

We do not intend a thorough study of point-set topology in this text, but only wished to provide a sampling. For more on the intricacies of this subfield of topology, you should consult a text devoted to it, such as James R. Munkres' *Topology: A First Course.*

3.4 Product spaces

In this section and the next, we investigate ways of creating new topological spaces from old ones. The euclidean plane $\mathbb{R}^2$ is an example of a product space.

(3.27) Definition. *Let X and Y be any spaces. The (cartesian) product of X and Y is the set of all ordered pairs (x, y); i.e.,*

$$X \times Y = \{(x, y) : x \in X, \, y \in Y\}$$

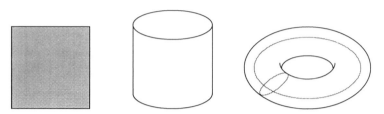

Fig. 3.4. Product spaces

Thus, $\mathbb{R}^2 = \{(x, y) : x, y \in \mathbb{R}\} = \mathbb{R} \times \mathbb{R}$. Similarly, if we let I denote the unit interval $[0, 1]$ and $\mathbb{S}^1$ the unit circle, then $I \times I$ is the unit square, $\mathbb{S}^1 \times I$ a cylinder, and $\mathbb{S}^1 \times \mathbb{S}^1$ the skin of a doughnut, or torus. (See Fig. 3.4.)

Given topologies on X and Y, a topology is naturally induced on $X \times Y$.

(3.28) Definition. *Let X be a topological space with topology $\mathfrak{T}$ and Y a topological space with topology $\mathfrak{T}'$. A basis for the product topology on $X \times Y$ is given by $\mathcal{B}$ where $N \in \mathcal{B}$ if N is of the form $N = O \times O'$ for $O \in \mathfrak{T}$ and $O' \in \mathfrak{T}'$. Projection functions $p_X : X \times Y \longrightarrow X$ and $p_Y : X \times Y \longrightarrow Y$ are defined by*

$$p_X(x, y) = x$$
$$p_Y(x, y) = y$$

Notice that one starts with the collection of all open sets on X and Y, but ends with a basis or system of neighborhoods on $X \times Y$, not the whole topology. We are not claiming that every open set in $X \times Y$ is of the form $O \times O'$. In $\mathbb{R}^2$, this definition gives us the basis of rectangular neighborhoods, which by Exercise 3.6 is equivalent to the standard topology on $\mathbb{R}^2$.

▷ **Exercise 3.28.** Show that the projection maps p_X and p_Y in Definition 3.28 are continuous.

▷ **Exercise 3.29.** Show that the product of two T_i-spaces is a T_i-space, for $i = 0, 1, 2$.

The product of two regular spaces is also regular, and the product of metric spaces is metric, but the proofs will be omitted. The product of two normal spaces need not be normal!

(3.29) Theorem. *Let X and Y be connected topological spaces. The product $X \times Y$ is connected.*

We first prove a lemma:

(3.30) Lemma. *Let A_α be a collection of connected topological spaces which have a point x in common (so $x \in A_\alpha$ for each α). Then $B = \bigcup_\alpha A_\alpha$ is connected.*

Proof. If B is not connected, then $B = C \cup D$, where C and D are non-empty disjoint open sets. The point x must be in either C or D; we will assume $x \in C$. Consider one of the sets A_α. Since A_α is connected, it must be contained in either C or D. Otherwise, $A_\alpha = \{C \cap A_\alpha\} \cup \{D \cap A_\alpha\}$ would show that A_α is not connected, since $C \cap A_\alpha$ and $D \cap A_\alpha$ are open in A_α. Since $x \in A_\alpha$ and $x \in C$, A_α must lie in C. This is true for each α, so $B = \cup A_\alpha \subseteq C$, and D must be empty. □

Proof of Theorem 3.29. Choose a point $(a, b) \in X \times Y$, and note that $\{a\} \times Y$ is homeomorphic to Y, and so $\{a\} \times Y$ is connected. Similarly, $X \times \{y\}$ is connected for each $y \in Y$. Consider the set formed by the union $B_y = (\{a\} \times Y) \cup (X \times \{y\})$. (See Fig. 3.5.)

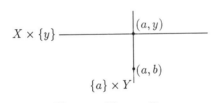

$$X \times \{y\} \underline{\hspace{3cm}} \begin{matrix} (a, y) \\ \\ (a, b) \end{matrix}$$

$$\{a\} \times Y$$

Fig. 3.5. The set B_y

This is connected by Lemma 3.30 since $\{a\} \times Y$ and $X \times \{y\}$ both contain the point (a, y). Also, note that the point $(a, b) \in B_y$, for each $y \in Y$. Now, form the union $\bigcup_{y \in Y} B_y$. This is connected by the lemma and is easily seen to equal $X \times Y$. □

(3.31) Theorem. *Let X and Y be compact topological spaces. The product $X \times Y$ is compact.*

Proof. Recall that a basis for $X \times Y$ is

$$\mathcal{B} = \{U \times V : U \text{ open in } X, V \text{ open in } Y\}$$

A preliminary lemma is necessary:

(3.32) Lemma. *A topological space X is compact if there exists a basis $\mathcal{B}$ for X such that every open cover of X which consists of elements of this basis has a finite subcover.*

▷ **Exercise 3.30.** Prove Lemma 3.32.

Proof of Theorem 3.31, continued. Let $\mathcal{O}$ be an open cover of $X \times Y$ such that every $O \in \mathcal{O}$ is of the form $O = U \times V$ where U is open in X and V is open in Y. Thus, $\mathcal{O}$ is a open cover consisting of basis sets, as specified in the lemma above. For each $y \in Y$, the set $X \times \{y\}$ is homeomorphic to X, so $X \times \{y\}$ is compact, and $\mathcal{O}$ also covers $X \times \{y\}$. Therefore, $\mathcal{O}$ has a finite subcover which covers $X \times \{y\}$; i.e., there exists sets $U_{y,i} \times V_{y,i}$ for $i = 1, 2, \ldots, n_y$ so that

$$X \times \{y\} \subseteq \bigcup_{i=1}^{n_y} (U_{y,i} \times V_{y,i})$$

as pictured in Figure 3.6.

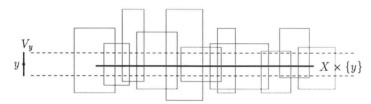

Fig. 3.6. The cover $U_{y,i} \times V_{y,i}$ for $X \times \{y\}$, and the set V_y

Consider $V_y = \bigcap_{i=1}^{n_y} V_{y,i}$ shown in Fig. 3.6. The set V_y is open in Y. Note that

$$X \times V_y \subseteq \left(\bigcup_{i=1}^{n_y} U_{y,i} \right) \times V_y = \bigcup_{i=1}^{n_y} (U_y \times V_{y,i}) \subseteq \bigcup_{i=1}^{n_y} (U_{y,i} \times V_{y,i})$$

since $V_y \subseteq V_{y,i}$ for each i. Form the set V_y for each $y \in Y$. The collection $\{V_y\}_{y \in Y}$ is an open cover for Y, and since Y is compact, there is a finite subcover $\{V_{y_j}\}_{j=1}^{m}$ for Y. Then

$$X \times Y \subseteq \left(\bigcup_{j=1}^{m} X \times V_{y_j} \right) = \bigcup_{j=1}^{m} (X \times V_{y_j}) \subseteq \bigcup_{j=1}^{m} \bigcup_{i=1}^{n_{y_j}} U_{y_j,i} \times V_{y_j,i}$$

Thus, $\mathcal{O}$ has a finite subcover and $X \times Y$ must be compact.   □

3.5 Quotient spaces

Another rich source of examples is the category of quotient or identification spaces.

(3.33) Definition. *Let X be a topological space with topology $\mathcal{T}$ and let $f : X \longrightarrow Y$ be a function onto a set Y. Define the quotient topology $\mathcal{T}'$ on Y by defining $U \subseteq Y$ to be open in Y if $f^{-1}(U)$ is open in X. Thus,*

$$U \in \mathcal{T}' \text{ iff } f^{-1}(U) \in \mathcal{T}$$

Note that we are not presupposing any topological structure on Y.

▷ **Exercise 3.31.** Show that the topology $\mathcal{T}'$ defined on Y in Definition 3.33 satisfies the conditions of Theorem 3.5.

▷ **Exercise 3.32.** Show that the function f of Definition 3.33 is continuous with respect to the topologies $\mathcal{T}$ and $\mathcal{T}'$.

An example of this construction is the function $f : \mathbb{R} \longrightarrow \mathbb{S}^1$ from the real numbers to the circle $\mathbb{S}^1 = \{(x, y) \in \mathbb{R}^2 : x^2 + y^2 = 1\}$ defined by

$$f(t) = (\cos(2\pi t), \sin(2\pi t))$$

Note that $f(t) \in \mathbb{S}^1$ since $\cos^2(2\pi t) + \sin^2(2\pi t) = 1$ for every $t \in \mathbb{R}$. Then $f(0) = (1,0)$, $f(\frac{1}{4}) = (0,1)$, $f(\frac{1}{2}) = (-1,0), \ldots$, $f(1) = (1,0)$, etc. The action of f may be pictured as wrapping the line counterclockwise around the circle, so that $f(0) = f(1) = \cdots = f(n) = (1,0)$ for $n \in \mathbb{Z}$ where $\mathbb{Z}$ denotes the integers.

Fig. 3.7. $\mathbb{R}$ with the identification defined by f wraps the line around the circle

By Definition 3.33, the open sets in $\mathbb{S}^1$ are the sets which have as inverse images the union of open intervals in $\mathbb{R}$ as pictured in Figure 3.7. This topology happens to coincide with the natural subspace topology on $\mathbb{S}^1$ as a subset of $\mathbb{R}^2$.

In practice, however, finding and writing out a function that does what you want is often rather dreary. For example, take a rectangular piece of paper and glue the left and right edges together to form a tube or cylinder, as in Figure 3.8. There is a function from the rectangle to the cylinder, and it is not hard to see where each point on the rectangle is sent on the cylinder by the function, but there is an easier way to describe the action: Leave the points on the inside of the rectangle and on the top and bottom edges alone, and glue the corresponding points on the left and right edges together. Of course, we need sensible rules for gluing before we do any more of this. The natural rules for gluing are:

(1) No point x can come unglued from itself.
(2) If x is glued to y, then y is glued to x.
(3) If x is glued to y and y is glued to z, then x is glued to z.

Doesn't that seem obvious? What we need is an equivalence relation!

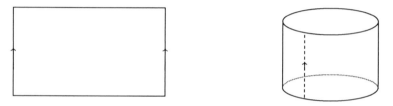

Fig. 3.8. Gluing a rectangle to make a cylinder

(3.34) Definition. *Let X be a topological space with $\sim$ an equivalence relation defined on X. Define the equivalence class of $x \in X$ by*

$$[x] = \{y \in X : y \sim x\}$$

The identification space $X/\sim$ is defined as the set of equivalence classes of the relation $\sim$, so

$$X/\sim = \{[x] : x \in X\}$$

The new space $X/\sim$ is just a fancy way of saying that a new space is created by taking the space X and gluing x to any y that satisfies $y \sim x$.

The cylinder is constructed in this manner from the unit square X in $\mathbb{R}^2$ where $X = \{(x,y) : 0 \leq x \leq 1, 0 \leq y \leq 1\}$ and x runs left to right and y down to up. Thus, the points on the bottom and top edges have $y = 0$ or 1, and the left and right edges have $x = 0, 1$. The equivalence relation is defined by letting the equivalence class $[(x,y)]$ consist only of the point (x,y) if $x \neq 0, 1$, so that the point (x,y) is not glued to any other point for points in the interior or the top and bottom edges of the square, and defining $(0, y) \sim (1, y)$ for each y. Points on the left edge of the square are glued to the corresponding points on the right edge. Thus,

$$[(x, y)] = \begin{cases} \{(x, y)\} & \text{if } x \neq 0, 1 \text{ and } 0 \leq y \leq 1 \\ \{(0, y) \sim (1, y)\} & \text{if } x = 0, 1 \text{ and } 0 \leq y \leq 1 \end{cases}$$

The resulting space $X/\sim$ looks like a cylinder, constructed from a square of paper by gluing the left and right edges together, as in Figure 3.8.

Another way of describing the example of Figure 3.7 is by defining an equivalence relation $\sim$ on $\mathbb{R}$ by $t \sim s$ if $s = t + n$ for some $n \in \mathbb{Z}$. Note that this is an equivalence relation since:

(1) Every $t \sim t$ because $t = t + 0$ and $0 \in \mathbb{Z}$.
(2) If $t \sim s$, then $s = t + n$ for $n \in \mathbb{Z}$. Since $t = s + (-n)$ and $-n \in \mathbb{Z}$, it follows that $s \sim t$.
(3) If $t \sim s$ and $s \sim r$, then $s = t + n$ and $r = s + m$ for $m, n \in \mathbb{Z}$. Then $r = (t + n) + m = t + (n + m)$. Since $(n + m) \in \mathbb{Z}$, $t \sim r$.

Then $\mathbb{R}/\sim$ is the set of real numbers with 0 glued to $1, 2, \ldots, -1, -2, \ldots,$ and $\frac{1}{4}$ glued to $\frac{5}{4}, \frac{9}{4}, \frac{13}{4}, \frac{17}{4}, \ldots,$ etc. In other words, take the interval $[0, 1]$ and glue the point 0 to 1 (thus creating a circle), then take the interval $(1, 2]$ and wrap this around the circle, gluing $\frac{5}{4}$ to the point $\frac{1}{4}$, $\frac{3}{2}$ to $\frac{1}{2}$, and 2 to 1, etc. Repeat for the interval $(2, 3]$, and so on. The space $\mathbb{R}/\sim$ looks like a circle $\mathbb{S}^1$, with the real line wrapped around it infinitely many times.

▷ **Exercise 3.33.** Describe the space $X/\sim$ for the following spaces and equivalence relations:

(1) Find $I/\sim$ for $X = I = [0, 1]$ and the equivalence classes defined by

$$[x] = \begin{cases} \{x\} & \text{if } 0 < x < 1 \\ 0 \sim 1 & \text{if } x = 0, 1 \end{cases}$$

(2) X is the unit square $\{(x, y) : 0 \leq x \leq 1, 0 \leq y \leq 1\}$ with the equivalence classes:

$$[(x, y)] = \begin{cases} \{(x, y)\} & \text{if } x \neq 0, 1 \text{ and } 0 \leq y \leq 1 \\ (0, y) \sim (1, 1 - y) & \text{if } x = 0, 1 \text{ and } 0 \leq y \leq 1 \end{cases}$$

(3) X is the disc $D^2 = \{(x, y) : x^2 + y^2 \leq 1\}$. Let $\mathbb{S}^1$ denote the boundary circle: $\mathbb{S}^1 = \{(x, y) : x^2 + y^2 = 1\}$. Define equivalence classes of a point (x, y) by

$$[(x, y)] = \begin{cases} \{(x, y)\} & \text{if } (x, y) \notin \mathbb{S}^1 \\ (x, y) \sim (1, 0) & \text{if } (x, y) \in \mathbb{S}^1 \end{cases}$$

(4) $X = \mathbb{S}^1 = \{(x, y) : x^2 + y^2 = 1\}$ and $(x, y) \sim (-x, -y)$ for each $(x, y) \in \mathbb{S}^1$.

Given a topological space X and an identification space $X/\sim$ constructed from X, there is a natural function $f : X \longrightarrow X/\sim$ defined by

$$f(x) = [x]$$

Give the set $X/\sim$ the quotient topology induced by X and f. This is referred to as the *identification topology* on $X/\sim$ and is a special case of Definition 3.33. In particular, a neighborhood of $[x] \in X/\sim$ is defined to be a set such that the inverse image in X is a union of neighborhoods of all points y with $y \sim x$.

In the example of the cylinder above, if the square is given the standard topology that it inherits as a subset of $\mathbb{R}^2$, so neighborhoods are discs intersected with the square, then the cylinder has neighborhoods that look like one of the sets illustrated in Figure 3.9:

(1) Small discs in the interior of the square (where no gluing took place) such as the point **x**.

(2) Half-discs (open in the relative topology) about points on the upper and lower edges, such as the point **z**.

(3) Since points $(0, y)$ and $(1, y)$ on the left and right edges of the square are identified, a neighborhood of $[(0, y)] = [(1, y)]$ consists of two half-discs (open in the relative topology on the square), centered at $(0, y)$ and $(1, y)$, joining to form a disc in the cylinder, centered at the point $\mathbf{y} = [(0, y)] = [(1, y)]$.

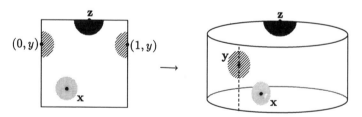

Fig. 3.9. Neighborhoods of points on the cylinder

▷ **Exercise 3.34.** Describe the neighborhoods of $X/\sim$ for each of the examples in Exercise 3.33, assuming that each X has the standard topology.

▷ **Exercise 3.35.** Show that if X is connected and $\sim$ is an equivalence relation on X, then $X/\sim$ is connected.

▷ **Exercise 3.36.** Show that if X is compact and $\sim$ is an equivalence relation on X, then $X/\sim$ is compact.

Chapter 4

Surfaces

"...did anyone know, he enquired, what a cubic foot of hardwood weighed? While all the women were respectfully waiting for the men to answer this, and all the men were knitting their brows, and trying to look as if they really did know, but had forgotten for the moment, Biddy suddenly said in a bewildered voice that the [tree] trunk was round and she thought cubes were square? Bruce replied quite civilly that cubes were cubic, and Biddy retorted that anyway, they weren't *round*, and she didn't see what square cubes had to do with a round trunk. Bruce, in a tone of slightly strained patience, pointed out that the trunk was not round, but cylindrical, and Biddy said that even if it was, all the cylinders she had ever seen were round, so what were they arguing about? This led to a general and extremely animated discussion (accompanied by much drawing of diagrams on the ground, and much quoting of such examples as pennies, wedding-rings, tennis balls, and water pipes)...from which it seemed to transpire, rather to everyone's astonishment, that simply nothing at all was round."

Eleanor Dark

4.1 Complexes

We now turn our attention to a more macroscopic view of spaces, and try to find more global and geometric descriptions of the objects of study. It is still our purpose to try to find topological invariants of the shape of a space, but the determination of whether two objects are topologically equivalent is an often difficult, indeed in many cases as yet unsolved, problem. In this text we will use what is called the *combinatorial* approach to topology, thus limiting ourselves to a large but relatively manageable group of objects. All the spaces we study will be built from a uniform set of building blocks. We then analyze the blocks themselves and how they are combined. The basic building blocks are called *cells* and are assembled into *complexes*. The cells are of varying dimensions, and so a complex has a stratified structure: a natural ordering of the cells by dimension. This text concentrates on spaces which are locally 2-dimensional.

(4.1) Definition. *An n-cell is a set whose interior is homeomorphic to the n-dimensional disc $D^n = \{\mathbf{x} \in \mathbb{R}^n : \|\mathbf{x}\| < 1\}$ with the additional property that its boundary or frontier must be divided into a finite number of lower-dimensional cells, called the faces of the n-cell. We write $\sigma < \tau$ if σ is a face of τ.*

    (0) *A 0-dimensional cell is a point A.*

    (1) *A 1-dimensional cell is a line segment $a = AB$, and $A < a$, $B < a$.*

    (2) *A 2-dimensional cell is a polygon (often a triangle), such as $\sigma = \triangle ABC$, and then AB, BC, $AC < \sigma$. Note that $A < AB < \sigma$, so $A < \sigma$.*

    (3) *A 3-dimensional cell is a solid polyhedron (often a tetrahedron), with polygons, edges, and vertices as faces.*

Note the faces of an *n*-cell are lower-dimensional cells: the endpoints of a 1-cell or edge are 0-cells, the boundary of a 2-cell or polygon consists of edges (1-cells) and vertices (0-cells), etc. (See Fig. 4.1.) These cells will be joined together to form complexes.

Fig. 4.1. 0-, 1-, and 2-cells

The space on the left in Figure 4.2 is not considered to be a cell, since its boundary is a circle which is not a 1-cell. It can easily be made into a 2-cell by dividing up the boundary into edges and vertices (1- and 0-cells), as illustrated on the right:

Fig. 4.2. The figure on the left is not a cell, but the one on the right is

The interior of an *n*-cell is topologically equivalent to the *n*-disc, D^n, but a cell has an additional structure dividing the points into interior points and boundary points, with the boundary points assigned to the facing cells.

Cells are glued together to form complexes, by gluing edge to edge and vertex to vertex and identifying higher-dimensional cells in a similar manner (though this requires care — we do not want to glue a rectangular face to a triangular face).

(4.2) Definition. *A complex K is a finite set of cells,*

$$K = \bigcup \{\sigma : \sigma \text{ is a cell}\}$$

such that:

(1) *if σ is a cell in K, then all faces of σ are elements of K;*
(2) *if σ and τ are cells in K, then $Int(\sigma) \cap Int(\tau) = \emptyset$.*

The dimension of K is the dimension of its highest-dimensional cell.

Condition (2) forbids intersections such as those of Figure 4.3.

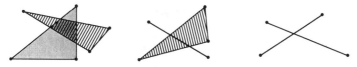

Fig. 4.3. Forbidden intersections

It must be emphasized that a complex is more than a set of points, since it also comes equipped with the structure given by the allotment of its points into cells of various dimensions. In each case above, notice that the intersections are homeomorphic to cells, but are not among the cells of the complex K.

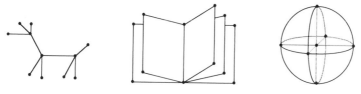

Fig. 4.4. Examples of complexes

Note that a topological object may be represented by many complexes, as in Figure 4.5.

Fig. 4.5. Several different complexes on the sphere

In Chapter 1, it was claimed that all of these are topologically equivalent, so these represent the same topological space, but with different complex structures.

Definitions 4.1 and 4.2 do not require that the boundary of a 2-cell consist of 1-cells, but only that the boundary be divided into lower-dimensional cells. Thus, another complex on the sphere is defined by taking a single 2-cell, and identifying its boundary to a point, rather like a draw-string purse.

Fig. 4.6. Another complex on the sphere

(4.3) Definition. *Let K be a complex. The set of all points in the cells of K is*

$$|K| = \{\mathbf{x} : \mathbf{x} \in \sigma \in K,\ \sigma\ \text{a cell in } K\}$$

is the space underlying the complex K, or the realization of K.

The distinction between a space and a complex is that K is a set of cells and $|K|$ is a set of points. A cell complex is a sort of layered structure, built up of cells of various dimensions.

(4.4) Definition. *Let K be a complex. The k-skeleton of K is*

$$K_k = \{k\text{-cells of } K\}$$

Note that K_k is a k-complex and $K = \bigcup_{k=1}^{n} K_k$ where $n = \dim(K)$.

▷ **Exercise 4.1.** Find the k-skeletons of the complexes in Figures 4.7 and 4.8, for $k = 0, 1, 2, 3$.

Fig. 4.7. A solid tetrahedron

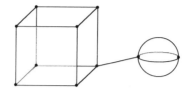

Fig. 4.8. A solid cube with a hollow sphere attached by an edge

A topological structure must be defined for complexes, by defining neighborhoods for all points in $|K|$. If K is a complex, then each cell in K has the standard topology on the n-dimensional ball as a subset of $\mathbb{R}^n$. We make a new topological space out of the union of cells by using the identification topology, defined for 2-complexes below. This is a detailed description of the topology discussed in Section 3.5, for the case of 2-complexes. First, we take a more careful look at how such complexes are formed and then define the neighborhoods. 2-complexes are built by taking a collection of polygons and identifying or gluing edges and vertices together. For example, take a square and glue a pair of opposite sides together to form a cylinder, and stick a triangular fin along the same edge. An extra edge (just to make things complicated) is glued to this figure at one endpoint to form the complex K pictured in Figure 4.9.

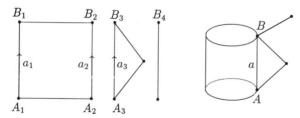

Fig. 4.9. The construction of complex K: glue all the edges marked a_i together, respecting the direction of the arrows, and identify the vertices accordingly

Thus, K is formed from two polygons (the square and the triangle) and the extra edge, by identifying the edges labeled a_1, a_2, and a_3 to form one new edge labeled a, and identifying the vertices A_1, A_2 and A_3 to the vertex A in K, and B_1, B_2, B_3, and B_4 to create the vertex B in K. The arrows can be used instead of labeling the vertices to specify a direction or orientation for the edges, and this direction must be respected when the gluing is done. Note that a single point on the edge a in K corresponds to three different points on the two polygons, whereas the vertex A corresponds to three vertices and B to four vertices.

2-complexes have three types of points: points lying in the interior of one of the 2-cells or polygons, points lying in the interior of one of the edges, and vertex points. We will define neighborhoods for each type of point separately. In general, we start with a set of polygons: $\mathcal{P} = \{P_i\}$.

Points lying inside a 2-cell: If a point x lies in the interior of a polygon P_i, then define a neighborhood of x to be any disc totally contained in the interior of P_i.

Points lying on an edge: If y lies on an edge b which is not a part of the boundary of any 2-cell, then define a neighborhood of y to be any open subinterval contained in the interior of b. If y is a point on edge $a \in K$, where a was formed from some collection of edges $a_1, \ldots, a_n$ chosen from the polygons $\{P_i\}$, then we may assume that this gluing was done in such a way as to respect the chosen direction of the edges. We first find all the points which joined to form the point y. Note that every edge, or 1-cell, is topologically equivalent to the unit interval $I = [0, 1]$, and so there are homeomorphisms from the edge a to I and from I to each edge a_i, chosen so that the initial points of the edges go to 0 and the terminal points go to 1. Let $f_i : a \longrightarrow a_i$ be the composite homeomorphism, for $i = 1, 2, \ldots, n$. (These functions are often easier to see than to write down.) Then the point y corresponds to the points $y_i = f_i(y)$, $i = 1, \ldots, n$. Each point y_i lies on the edge a_i in some polygon P_k and has a half-disc neighborhood in that polygon.

Fig. 4.10. A neighborhood of a point y lying in the interior of a 1-cell

These half-discs can be chosen so that they do not overlap each other, do not include any vertices, and have matching radii. When the edges $a_1, \ldots, a_n$ are glued together to form the edge a, the half-discs are also glued together along their diameters to create a neighborhood of the edge point y in the complex K, as in Figure 4.10.

Vertex points: Let B be a vertex in the complex K, formed by identifying vertices $B_1, B_2, \ldots, B_m$ from the set of polygons and edges. Each B_i is either in one of the polygons $P_k \in \mathcal{P}$, or B_i could belong to an edge b which is not part of one of the polygons (as in the vertex B_4 in the example above). If B_i lies in a polygon P_k, choose a relative neighborhood of B_i in the polygon and note that this will look like a sector of a disc. These disc sectors must be chosen so that they do not overlap one another, do not intersect any other vertices, and have matching radii. If B_i lies on an edge only, then a relative neighborhood of B_i is a half-open interval with B_i at the end. After building the complex K, these sectors and half-open intervals are joined to form a neighborhood of B. (See Fig. 4.11.)

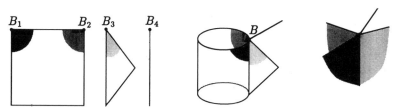

Fig. 4.11. A neighborhood of a vertex point B

Note that the neighborhoods of points in complexes in this topology are not always discs. However, in this text the topology defined as above will coincide with the relative topology if the figure is imagined as a subset of $\mathbb{R}^n$. The definitions for continuous functions, connected sets, and compact sets follow as in Chapter 2 and 3.

It is usually easier for those of us with no artistic talent to give directions for the construction of complexes from its cells than to attempt a drawing or to list the cells to be identified. For example, a cylinder can be constructed from a rectangle by gluing the edges together. We mark two opposite edges of the rectangle with arrows to indicate instructions to glue those sides together in the direction of the arrows to obtain a cylinder. This rectangle provides a convenient way of representing the cylinder. Notice that the cell structure of the cylinder can be seen both in the rectangular schematic drawing and in the assembled cylinder as pictured in Figure 4.12. The rectangle representation has another advantage over the picture: One can see the whole structure, whereas in the picture the part of the cylinder in the back is obscured. The disadvantage to this representation is that one must remember that the sides with the arrows must be considered as being glued together. One must try to bear in mind that those sides are, in effect, two copies of a single 1-cell; i.e., the arrows indicate that the marked sides are really to be considered as one. This type of representation is called a *planar diagram*.

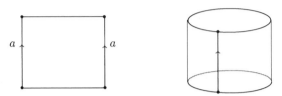

Fig. 4.12. A complex on the cylinder

Another topological object is obtained from identifying the opposite sides of a square in pairs. Bend and glue both pairs of sides to obtain an inner tube shape or the surface of a doughnut (mentioned in Section 3.4 as $\mathbb{S}^1 \times \mathbb{S}^1$) known in mathematical circles as a *torus* and denoted by $\mathbb{T}^2$. (See Fig. 4.13.)

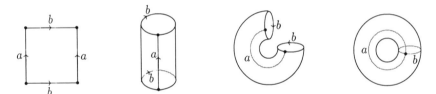

Fig. 4.13. Planar diagram for the torus $\mathbb{T}^2$

A disc with a zipper gives a football shape, topologically equivalent to the sphere, $\mathbb{S}^2$. (See Fig. 4.14.)

Fig. 4.14. Planar diagram for the sphere $\mathbb{S}^2$

There are, of course, many different planar diagrams for any surface, just as there are many complexes representing the surface.

Gluing one pair of edges of a rectangle with a twist gives a Möbius band as in Figure 4.15. Compare this with the planar diagram of the cylinder.

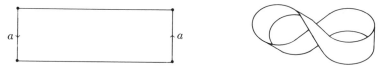

Fig. 4.15. Planar diagram for the Möbius band

The Möbius band has the interesting property of having only one side, in contrast to the cylinder. It is easy to imagine a cylinder with the outside painted one color and the inside another. Try painting a Möbius strip.

Another peculiarity of the Möbius band occurs when one cuts it along the dotted line (called the meridian), and then follows the gluing instructions of the edges, as in Figure 4.16. The end result is a planar diagram for a cylinder. However, if one actually constructs a Möbius band of paper and cuts it as described above, one gets something that looks like a cylinder with two twists.

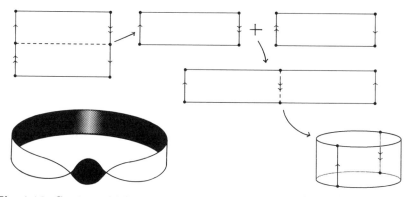

Fig. 4.16. Cutting a Möbius band along the meridian, in theory and in practice

This leads to a short digression on intrinsic versus extrinsic properties. Intrinsic properties have to do with the object itself, in contrast to extrinsic properties which describe how the object is embedded in the surrounding space. The cylinder and the band with two twists are intrinsically the same:

There is a homeomorphism between them. The difference between them lies in how they sit in our 3-dimensional universe. Both are assembled from a rectangle by the same gluing instructions but one is given two twists before gluing. We shall, for the greater part of this course, concern ourselves only with intrinsic properties. An object exists in itself, regardless of whether or how it may be embedded inside a larger space.

▷ **Exercise 4.2.**   Cut a Möbius strip in thirds as above. What do you get? What if it is cut in fourths or fifths? Try to invent a general rule.

▷ **Exercise 4.3.**   All but one of the spaces in Figure 4.17 are intrinsically the same:

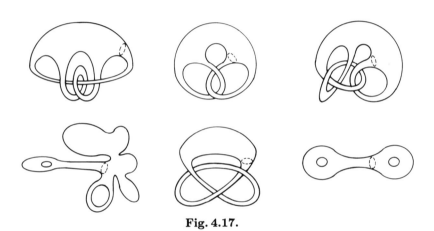

Fig. 4.17.

4.2 Surfaces

The complexes that have been studied most intensely by topologists are the *manifolds*. This text is devoted to the study of the 2-dimensional ones.

(4.5) Definition. *An n-dimensional manifold is a topological space such that every point has a neighborhood topologically equivalent to an n-dimensional open disc with center $\mathbf{x}$ and radius r, $D^n(\mathbf{x}, r) = \{\mathbf{y} \in \mathbb{R}^n : \|\mathbf{x} - \mathbf{y}\| < r\}$. We further require that any two distinct points have disjoint neighborhoods. A 2-manifold is often called a surface.*

The second condition of this definition, which says that manifolds are Hausdorff (Definition 3.22), is needed for technical reasons: There are certain pathological spaces which have the local disc-like property but are not

Hausdorff. We will not deal with such spaces in this text. Just hope you never meet one on a dark night.

An n-manifold is locally n-dimensional. If one imagines oneself as a very small, rather nearsighted, astigmatic bug living on a 2-manifold, one would not see any difference between what the manifold looks like and what the plane would look like. If one were also a stay-at-home sort of bug, one would never find out that one was not living on the plane. Of course, a more adventurous bug would find out the difference between the sphere and the plane (remember Magellan), or the torus and the sphere.

▷ **Exercise 4.4.** Figure out ways that a rather handicapped but adventurous bug could find out if his planet were $\mathbb{R}^2$, $\mathbb{S}^2$, $\mathbb{T}^2$, or the Möbius band.

The sphere, denoted $\mathbb{S}^2$, is a surface, even though it exists in a 3-dimensional background. If one considers a point $\mathbf{x} \in \mathbb{S}^2$ as a point in $\mathbb{R}^3$, of course it has a neighborhood that looks like a ball. As a point on the sphere, with the relative topology of Definition 2.13, $\mathbf{x}$ has neighborhoods of the form

$$N = \{\mathbf{y} \in \mathbb{R}^3 : \|\mathbf{x} - \mathbf{y}\| < r \text{ and } \mathbf{y} \in \mathbb{S}^2\}$$

These neighborhoods look like 2-dimensional discs which have been warped a bit. The sphere is represented by a disc with a zipper. On this planar diagram for the sphere, neighborhoods of interior points like $\mathbf{x}$ are easy to illustrate. Points along the edge, such as $\mathbf{y}$, have neighborhoods as pictured in Figure 4.18. When the edges of the planar diagram are glued (or sewn or stapled) together, these half-discs are also glued to form a neighborhood of $\mathbf{y}$ topologically equivalent to a disc.

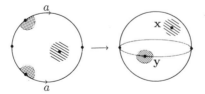

Fig. 4.18. Neighborhoods of $\mathbf{x}$ and $\mathbf{y}$ in $\mathbb{S}^2$

Similarly, the torus, if embedded in $\mathbb{R}^3$, can be seen to have warped disc-like neighborhoods. In building the torus $\mathbb{T}^2$ from the standard planar diagram, interior points have obvious disc neighborhoods. Edge points will have neighborhoods assembled from two half-discs bordering the edges of the square planar diagram. The four corners of the square end up being glued together to form a single vertex in $\mathbb{T}^2$. This vertex has a neighborhood represented in Figure 4.19. The four quarter-circles form a disc, when identified by the directions for assembling the torus.

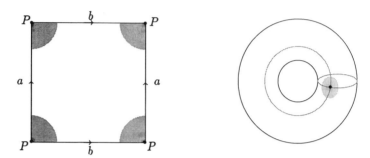

Fig. 4.19. A neighborhood of a vertex point on the torus

The cylinder and the Möbius band are not quite surfaces, since they have points with only half-disc neighborhoods.

(4.6) Definition. *An n-manifold with boundary is a topological space such that every point has a neighborhood topologically equivalent to either an open n-dimensional disc or the half-disc*

$$D^n_+ = \{\mathbf{x} = (x_1, x_2, \ldots, x_n 0) \in \mathbb{R}^n : \|\mathbf{x}\| < r \text{ and } x_n \geq 0\}$$

Points with half-disc neighborhoods are called boundary points. (See Fig. 4.20.)

Fig. 4.20. Boundary point and neighborhood

The boundary of the cylinder consists of two circles, and the boundary of the Möbius band is a single circle. The cylinder and the Möbius band have boundary edges, but in the sphere and torus all the edges are matched with other ones, so the complex closes on itself and encloses a cavity. In classifying the surfaces, the ability to enclose a cavity will turn out to be a distinguishing feature.

If one takes a square and glues the edges together in a different manner, one gets the surface known as the *Klein bottle*, denoted $\mathbb{K}^2$, of Figure 4.21. One has to pass the neck of the bottle through the side (without touching the side) in order to match up the arrows. This cannot be done in a 3-dimensional world, but is theoretically possible in a 4-dimensional universe.

Fig. 4.21. The Klein bottle, $\mathbb{K}^2$

Think of trying to connect the point A to point B without crossing the (infinite) line in Figure 4.22. This cannot be done on the plane; there simply is not enough room. However, in three dimensions, it is easy to hop over the line to join A to B. The same sort of argument allows one to construct a Klein bottle in $\mathbb{R}^4$. The Klein bottle has no boundary but does not enclose a cavity (look at the illustration in Figure 4.21 and imagine a Klein bottle filled with liquid).

Fig. 4.22. Get from A to B without crossing the line

Another property that differentiates the Klein bottle from the torus is the fact that the Klein bottle contains a Möbius band. Indeed, the Möbius band inside the Klein bottle seems to be why the Klein bottle does not enclose a cavity. Look at Figures 4.21 and 4.23 and trace the Möbius band. Any fluid in the bottle would run along the Möbius band and flow from inside the bottle to outside, because the Möbius band is a one-sided complex.

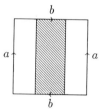

Fig. 4.23. There is a Möbius band inside the Klein bottle.

(4.7) Definition. *A non-orientable surface is one which contains a Möbius band.*

One would, ideally, like to classify all complexes; i.e., list all possible types up to homeomorphism, and describe their characteristics. That is entirely too complex a task so instead we will classify the surfaces only.

Consider the characteristics of the examples above: the cylinder and the Möbius band have boundary, but the others do not; the sphere and the torus both enclose cavities; the torus and the Klein bottle both have sections that look like Figure 4.24, which we shall describe, appropriately enough, as a handle.

Fig. 4.24. A handle

So far, we have considered a number of surfaces and their characteristics. All of the characteristics of Table 4.1 are topological properties. But are these enough to completely characterize a surface? In other words, if one knows that S is an orientable handleless surface with no boundary which encloses a cavity, does it follow that S must be a sphere? (The answer to this question is yes.)

Table 4.1. Surface characteristics

Surface	boundary	cavity	handle	orientable
Sphere	no	yes	no	yes
Torus	no	yes	yes	yes
Cylinder	yes	no	no	yes
Mobius Band	yes	no	no	no
Klein Bottle	no	no	yes	no

Have we listed all possible surfaces? No, since we have forgotten the manifold in Figure 4.25, which is called a two-handled torus. One can construct tori with any number of handles.

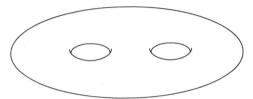

Fig. 4.25. A two-handled torus, $2\mathbb{T}^2$

(4.8) Definition. *Let S_1 and S_2 be two surfaces. Remove a small disc from each of S_1 and S_2, and glue the boundary circles of these discs together to form a new surface called the connected sum of S_1 and S_2, written $S_1 \# S_2$. (See Fig. 4.26.)*

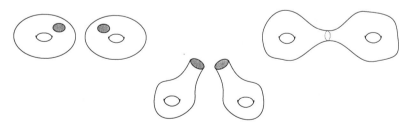

Fig. 4.26. The connected sum construction

This operation can also be pictured on the planar diagrams. Let us construct $\mathbb{T}^2 \# \mathbb{T}^2$, illustrated in Figure 4.27. The little holes can be made anywhere, and it is somewhat more efficient to place them at the vertices.

Stretch the holes open and glue together. The resulting figure is the planar diagram of the the 2-handled torus.

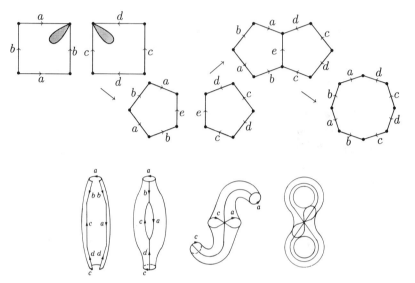

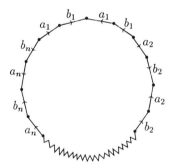

Fig. 4.27. The connected sum of two tori, $\mathbb{T}^2 \# \mathbb{T}^2$

The planar diagram for the n-handled torus is shown in Figure 4.28.

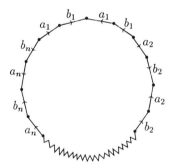

Fig. 4.28. The planar diagram of an n-handled torus, $n\mathbb{T}^2$

One is suddenly faced with many more examples of surfaces: n-handled tori, the connected sum of tori with Klein bottles, or Klein bottles with Klein bottles. Indeed, any even-sided polygon with the edges identified in pairs is a planar diagram for some surface.

Consider the planar diagram in Figure 4.29.

Fig. 4.29. The projective plane, $\mathbb{P}^2$

It is a diagram for some surface with neighborhoods at interior points obviously discs and at edge points, two half-discs glued together to form a disc. Think about trying to assemble this figure. This is called the *projective plane*, denoted $\mathbb{P}^2$, and cannot exist in 3-dimensional space. Notice that there is a Möbius band inside $\mathbb{P}^2$, so this surface is non-orientable.

▷ **Exercise 4.5.** Prove that $\mathbb{P}^2$ is a sphere $\mathbb{S}^2$ with a disc removed and a Möbius band glued in its place.

Consider the connected sum of two projective planes, $\mathbb{P}^2 \# \mathbb{P}^2$, as pictured in Figure 4.30. This can be rearranged to prove that $\mathbb{P}^2 \# \mathbb{P}^2 = \mathbb{K}^2$.

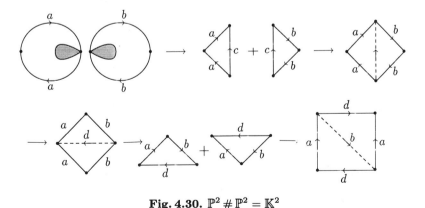

Fig. 4.30. $\mathbb{P}^2 \# \mathbb{P}^2 = \mathbb{K}^2$

It may strike you as odd that I am suddenly cutting and pasting like mad, in spite of the proviso made earlier that cutting and pasting are not homeomorphisms. Any cut can be repaired by gluing things back just the way they were. The last diagram of Figure 4.30 contains a promise to glue the d edges back together. Another way around this difficulty is to recognize that by the conventions we are using no cut has really been made, since by labeling two edges with d we have already (theoretically) glued them together in the identification topology. The planar diagrams provide a different picture of the space, not a different space. Thus, dissections as in Figure 4.30 represent homeomorphisms.

The planar diagram of the connected sum of n projective planes is pictured in Figure 4.31.

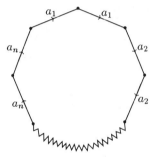

Fig. 4.31. The connected sum of n projective planes, $n\mathbb{P}^2$

▷ **Exercise 4.6.** Are the surfaces in Figure 4.32 topologically equivalent?

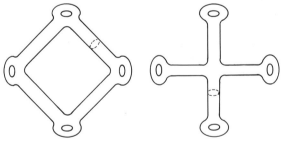

Fig. 4.32.

▷ **Exercise 4.7.** Prove that the connected sum of two surfaces is a surface, and that for any surface S, $S\#\mathbb{S}^2 = S$.

▷ **Exercise 4.8.** What figure does one obtain from a Möbius band if one shrinks the boundary circle to a point?

4.3 Triangulations

We shall return to the problem of classifying all surfaces in Section 4.4, after a more careful study of how they can be put together. It is often an advantage to use only triangular cells in a complex, so that one has only one type of n-cell. Then one does not have to worry about how to glue a triangle to a square, or whether the labeling $ABCD$ stands for a rectangle or a tetrahedron (see Fig. 4.33). Triangular objects (vertices, edges, triangles, tetrahedra, etc.) have the nice property that a triangular n-cell will have $(n+1)$ vertices, so that the number of vertices identifies the dimension of the cell.

Fig. 4.33. $ABCD$ could designate either a tetrahedron or a quadrilateral

(4.9) Definition. *A locally 2-dimensional topological space X is triangulable if a 2-complex structure K can be found with $X = |K|$ and K has only triangular cells satisfying the additional condition that any two triangles are identified along a single edge or at a single vertex or are disjoint. A triangulated complex K is called a simplicial complex or a triangulation on X. A cell of a simplicial complex is called a simplex.*

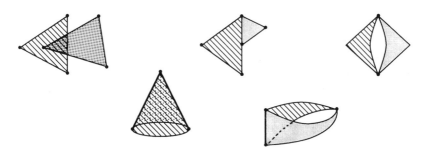

Fig. 4.34. These are not triangulations

This is an extension of Definition 4.2, with triangles the only allowed 2-cell. Note that Definition 4.9 implies more than that the space X is divided up into triangles. Several of the examples of Figure 4.34 are valid cell complexes with only triangular 2-cells, but are not triangulations.

Definition 4.9 implies not only that the complex will be divided nicely into triangles (the simplest polygons) but, as a valuable by-product of the added condition, that each vertex, edge, and triangle can be uniquely labeled by assigning a letter to each vertex. Simplicial complexes have the advantage of only requiring the position of the vertices to determine the position and dimension of the cell, but the disadvantage of usually requiring more cells than a general cell complex. In general, one works with cell complexes whenever possible, but falls back on simplicial complexes when more control is needed or for notational convenience in the proofs.

▷ **Exercise 4.9.**   Explain exactly how each of the complexes in Figure 4.34 violates Definition 4.9 and explain why each gives an ambiguous labeling.

We often want to triangulate a given complex. Each face of a 2-complex is a polygon and can easily be divided into triangles by introducing a new vertex in the interior of the polygon and then connecting this vertex to each of the vertices on the boundary, as in Figure 4.35.

Fig. 4.35. Dividing a polygon into triangles

However, it should be noted that although this process gives a method for dividing any 2-complex into triangles, it does not always give a triangulation satisfying Definition 4.9. The complex of Figure 4.36 has two different triangles labeled PQR, and so cannot be not simplicial. The triangles must be further subdivided.

Fig. 4.36. The planar diagram for the sphere is divided into triangles, but this is not a triangulation

One uniform way of subdividing triangles is:

(4.10) Definition. *K is a 2-complex with triangular 2-cells. A new complex K' called the barycentric subdivision of K is formed by introducing a new vertex at the center of each triangle and a new vertex at the midpoint of each edge and drawing edges from the center vertex to each of the new midpoint vertices and to the original vertices. In general, this is described as creating a new vertex v_σ in the center of every cell σ in K, including any vertex P when we define $v_P = P$, and add a connecting cell from v_σ to v_τ whenever $\sigma < \tau$. (See Fig. 4.37.)*

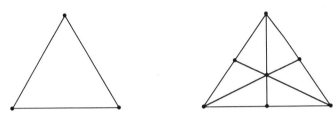

Fig. 4.37. The barycentric subdivision of a 2-cell

For the sphere in Figure 4.36, the barycentric subdivision of the complex gives a triangulation satisfying Definition 4.9, illustrated in Figure 4.38.

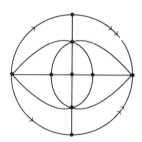

Fig. 4.38. A triangulation on the sphere

The barycentric subdivision may not always give a triangulation of a 2-complex with triangular faces. The second barycentric subdivision always suffices and is illustrated in Figure 4.39, but this proof is rather messy and so is omitted.

In practice, however, it is more efficient to only subdivide those edges and triangles which conflict with the conditions of Definition 4.9, thus minimizing the number of new triangles.

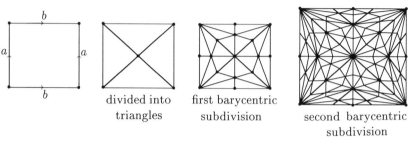

divided into   first barycentric
triangles   subdivision

second barycentric
subdivision

Fig. 4.39. Triangulating the torus by using the barycentric subdivision twice

▷ **Exercise 4.10.** Find a triangulation for $\mathbb{P}^2$.

We may now redefine surfaces in terms of their combinatorial structure:

(4.11) Definition. *A triangulated surface (without boundary) is a simplicial 2-complex such that:*

(1) *each edge is identified to exactly one other edge;*
(2) *the triangles meeting at a vertex can be labeled $T_1, T_2, \ldots, T_n$ with adjacent triangles in this sequence identified along an edge and T_n glued to T_1 along an edge.*

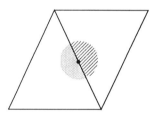

Fig. 4.40. An edge point on a surface

Fig. 4.41. An edge point on a complex which is not a surface

Condition (1) guarantees points on an edge belong to exactly two triangles, and so a disc-like neighborhood exists for each point on the edge resulting from gluing two half-discs together, one from each triangle, as illustrated in Figure 4.40. Edge points of surfaces cannot have neighborhoods as in Figure 4.41.

Condition (2) ensures that a neighborhood at a vertex looks like the disc as in Figure 4.42 and eliminates the possibility of neighborhoods as pictured in Figure 4.43.

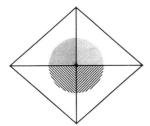

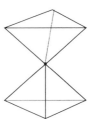

Fig. 4.42. A vertex point on surface

Fig. 4.43. A vertex point on a complex which is not a surface

Rado (1925) proved that every compact surface can be triangulated, so we may assume that all surfaces are equipped with a triangulated cell structure.

(4.12) Theorem. *A surface is compact if and only if any triangulation uses a finite number of triangles.*

Proof. If S is a compact surface, then let us see what happens if we assume S has a simplicial complex with an infinite number of triangles. Since S is a combinatorial surface, by Definition 4.11(2) there are only finitely many triangles meeting at each vertex. Therefore, if there are infinitely many triangles, then there must be an infinite number of vertices which can be considered as a sequence $\{v_i\}_{i=1}^{\infty}$. Since S is compact, Definition 2.23 implies that this sequence has a limit point $v \in S$. If v is in the interior of some triangle, then v has a neighborhood totally inside this triangle, but this contradicts v being a limit point of a set of vertices. If v is along some edge, then v has a neighborhood taken from two triangles, which will not contain any vertices, so again there is a contradiction. If v is a vertex, then v is a vertex of triangles $T_1, T_2, \ldots, T_n$ and has a neighborhood composed of sectors of these triangles and this neighborhood contains no other vertex. Thus, there is no other vertex in a neighborhood of v. This contradiction implies that there are only finitely many vertices and, therefore, finitely many triangles.

If S is a surface built from finitely many triangles, we must show that S is compact. Let $\{x_i\}_{i=1}^{\infty}$ be a sequence of points in S. We must show that the sequence has a limit point in S. Since there are only finitely many triangles and infinitely many points in the sequence, some triangle T contains an infinite number of the points. Let $\{x_j'\}_{j=1}^{\infty}$ be the subsequence of points in T. A triangle is a closed bounded subset of $\mathbb{R}^2$, so by the Heine-Borel Theorem 2.24, T is compact. Thus, the subsequence $\{x_j'\}_{j=1}^{\infty}$ has a limit point $x \in T \subseteq S$. By Definition 2.10, x is also a limit point for the sequence $\{x_i\}_{i=1}^{\infty}$. Thus, S is compact. $\square$

As an example of an infinite simplicial complex on a noncompact surface, consider the punctured disc, $D^2(\mathbf{0}, 1) - \{\mathbf{0}\}$, which is not closed and so is not compact, illustrated in Figure 4.44.

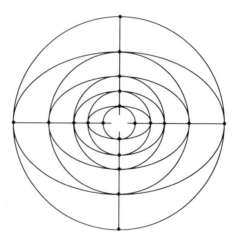

Fig. 4.44. An infinite simplicial complex on $D^2(\mathbf{0}, 1)$, where the rings have radius $1, \frac{1}{2}, \frac{1}{3}, \frac{1}{4}, \ldots$

(4.13) Theorem. *A surface is connected if and only if a triangulation can be arranged in order $T_1, T_2, \ldots, T_n$ with each triangle having at least one edge identified to an edge of a triangle listed earlier.*

Proof. Let S be a connected surface with a triangulation. Choose T_1 to be any triangle. Since any triangle is connected, T_1 is connected. Since S is a connected surface, there is a triangle T_2 which is glued to T_1 along an edge, forming a connected complex with two triangles. Then choose a triangle T_3 glued along an edge to the complex formed by $T_1 \bigcup_{\text{glue}} T_2$, so $T_1 \bigcup_{\text{glue}} T_2 \bigcup_{\text{glue}} T_3$ must be connected, etc. If S is connected, this process can be continued as long as there are free edges and triangles left.

If S is not connected, then S will have at least two connected components. Let $T_1, T_2, \ldots, T_k$ be the triangles in the first component of S and $T_{k+1}, T_{k+2}, \ldots, T_m$ the triangles comprising the second component. Note that T_{k+1} is not glued to any of the triangles $T_1, T_2, \ldots, T_k$, since these lie in different connected components.   □

▷ **Exercise 4.11.** Show that the connected sum of two triangulable spaces is triangulable.

▷ **Exercise 4.12.** Which of the following are simplicial complexes on surfaces?

   (1) ABD, BCD, ACD, ABE, BCE, ACE.
   (2) ABE, BCE, CDE, ABF, BCF, ADF, AEF, ADE.
   (3) ABC, BCD, CDE, ADE, ABE, CEF, BEF, BDF, ADF, ACF.
   (4) ABD, EFG, BCD, EGH, ABC, EFH, ACD, FGH.
   (5) ABD, BDE, ACD, CDF, BCF, BEF, ABG, BGH, CHI, ACI, BCH, AGI, EGH, EFH, FHI, DFI, DGI, DEG.
   (6) ABE, AEF, ABF, BCG, BEG, CGD, CGJ, GEH, GHJ, JHK, HIK, DIK, DIC, ADJ, AJK, AKD, BIC, BIF, FHI, EFH

4.4 Classification of surfaces

In this section we completely categorize (up to topological equivalence) all compact surfaces without boundary.

(4.14) Theorem. *Every compact connected surface is homeomorphic to a sphere, a connected sum of n tori, or a connected sum of n projective planes.*

Proof. The proof is constructive — not only is the result proved, but a method is outlined by which any surface can be reduced to one of the chosen forms. Assume that S is represented by a triangulated complex.

Step 1: Build a planar model of the surface.

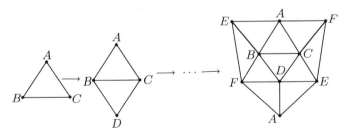

Fig. 4.45. Building the planar model

Since S is a compact surface, there is a simplicial complex on S with only finitely many triangles. Since S is connected, Theorem 4.13 implies that the list of triangles can be rearranged so that each triangle is glued to an earlier one. For example, if we are given the list of triangles: ABC, BDF, CEF, DAF, EBF, CDE, BCD, DEA, EAB, ACF, we reorder them as ABC, BCD, BDF, DAF, EBF, CDE, CEF, DEA, EAB, ACF. There are many such reorderings possible. Then assemble these in the chosen order to form a polygon representing a planar diagram for the surface as in Fig. 4.45. There are usually several ways to do this step, but the process of this theorem will reduce whichever planar diagram you start with to a standard form.

Of course, the outer edges must still be glued to form the surface. Since S is a surface, each edge is identified to exactly one other edge. The inside edges are already identified, so the external edges are identified in pairs. At this point, I usually relabel the edges marked AB with single letters such as a, since one no longer has to worry about the inside identifications. In the example above we get a hexagon with sides labeled as in Figure 4.46.

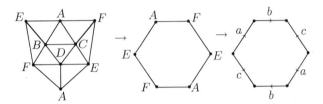

Fig. 4.46. Relabel the model: $a = EA$, $b = AF$, $c = EF$

Step 2: A Shortcut.

Throughout this theorem, diagrams are drawn with jagged edges to represent any unspecified edges that are not under current study.

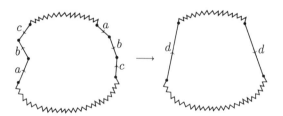

Fig. 4.47. Simplify by labeling $abc^{-1} = d$

The length of the procedure outlined in this theorem can be shortened by using a further trick whenever possible: If a string of edges occurs twice in exactly the same order, taking into account the directions of the edges, we can relabel to consider the string as a single edge. For example, relabel abc^{-1} as d in Figure 4.47. Note that in identifying the string of edges marked abc^{-1}, we get the same surface as when we glue the edges labeled d together, so both planar diagrams of Figure 4.47 give the same surface. This shortcut should be used whenever possible, so watch for it when going through the other steps of this algorithm.

Note that edges can occur in two forms: *opposing pairs* or *twisted pairs*. If one travels around the perimeter counterclockwise, a twisted pair (Fig. 4.48) will point the same direction, and an opposing pair (Fig. 4.49) in opposite directions.

Fig. 4.48. Opposing pair

Fig. 4.49. Twisted pair

Step 3: Eliminate adjacent opposing pairs.

Adjacent opposing pairs can be eliminated by folding them in and gluing the edges together, as in Figure 4.50.

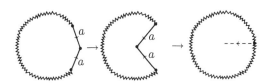

Fig. 4.50. Eliminate adjacent opposing pairs of edges

Note that if the only edges form an adjacent opposing pair, then the surface S is the sphere $\mathbb{S}^2$. Otherwise, proceed to Step 4.

Step 4: Eliminate all but one vertex

First, figure out how many vertices there are in the planar diagram. The surface in Figure 4.51 has three different vertices P, Q, R. Vertex P is the initial point of c and the terminal point of a. Vertex Q is the tail of both a and b; R is the head of both b and c.

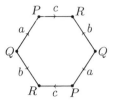

Fig. 4.51. Three different vertices

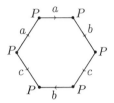

Fig. 4.52. Only one vertex

The planar diagram of Figure 4.52 has only one vertex, since $P =$ the head of a, which is the same as the tail of a, which is glued to the tail of c, which is glued to the head of b, which is glued to the head of c, which is the tail of b.

Now choose which vertex you fancy most. In the process depicted in Figure 4.53, we want to change the Q vertex into a P vertex. Cut along the dotted line and reglue along b, the edge which contains the undesirable vertex Q, to obtain one more P vertex and one less Q vertex.

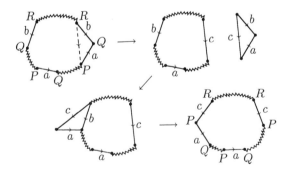

Fig. 4.53. Eliminating a Q vertex in favor of a P vertex

Note that if the triangle had been glued back along the edge a instead of b, then we would have the same number of P vertices as before, one less Q vertex and one more R vertex. Repeat Step 4 as necessary to eliminate all but one vertex.

Step 5: Collecting twisted pairs.

A twisted pair of edges labeled a may be made adjacent by cutting along the dotted line and regluing along the original edge a, as in Figure 4.54.

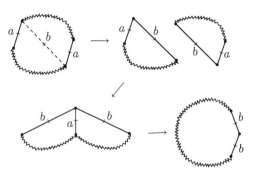

Fig. 4.54. Bringing twisted pairs together

If, at the completion of this step as often as needed, there are no edges of any other type, then the surface is a connected sum of $n\mathbb{P}^2$'s.

Step 6: Collecting pairs of opposing pairs.

If Steps 1 through 5 have been performed, then any opposing pairs must occur in pairs (rather a confusing phrase), as in Figure 4.55.

Fig. 4.55. Pairs of opposing pairs

Otherwise, we would have something with only one pair of opposing edges, and all other edges must occur as twisted pairs and will be adjacent since Step 5 has already been carried out.

For example, in Figure 4.56, count the vertices. Note that $P = $ head of $a = $ tail of b and $Q = $ tail of $a = $ head of $d = $ tail of d. Thus, we have more than one vertex, contradicting the assumption that Step 4 has been

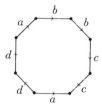

Fig. 4.56. Only one pair of opposing edges

completed. The only way that both the head and the tail of a can signify the same vertex is if there is another pair of edges divided by the a's. Since by Step 5 all twisted pairs have already been made adjacent, this other pair of edges must be opposing. This argument is clearly contingent on the steps of this theorem being done in order.

Pairs of opposing pairs a and b may be brought together by first cutting along the dotted line c connecting the heads of the a edges, regluing along b, then cutting along the dotted line d connecting the heads of c, and regluing along a, to to obtain a toroidal pairing $cdc^{-1}d^{-1}$ as in Figure 4.57.

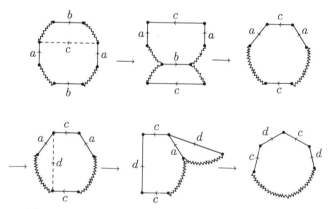

Fig. 4.57. Bringing pairs of opposing pairs together

If, at the completion of this step as often as needed, there are no edges of twisted type, then the surface is a connected sum of $n\mathbb{T}^2$'s.

Step 7: Combinations of $\mathbb{T}^2$ and $\mathbb{P}^2$.

At this point, either the surface has been classified or the planar diagram has both twisted pairs and toroidal pairings, so that the surface is a connected sum of projective planes and tori. This case is reduced by using the following lemma.

(4.15) Lemma. $\mathbb{T}^2 \# \mathbb{P}^2 = \mathbb{P}^2 \# \mathbb{P}^2 \# \mathbb{P}^2.$

Proof. See Figure 4.58.

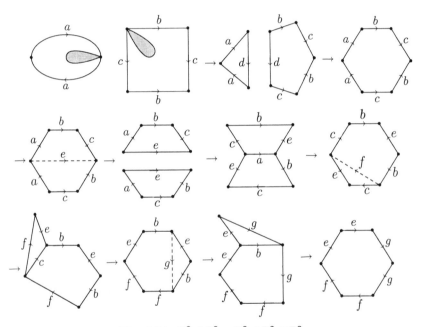

Fig. 4.58. $\mathbb{T}^2 \# \mathbb{P}^2 = \mathbb{P}^2 \# \mathbb{P}^2 \# \mathbb{P}^2$

□

Proof of Theorem 4.14, continued. Thus, a combination of tori and projective planes can be converted to a connected sum of projective planes only.

It must be stressed that the steps of Theorem 4.14 must be done in order, except that Steps 2 and 3 can and should be applied whenever possible. □

▷ **Exercise 4.13.** Classify the surfaces from Exercise 4.12.

▷ **Exercise 4.14.** How is $\mathbb{T}^2 \# \mathbb{K}^2$ classified? $\mathbb{K}^2 \# \mathbb{K}^2$?

▷ **Exercise 4.15.** Let $S_1 = m\mathbb{T}^2$ (the connected sum of m tori) and $S_2 = n\mathbb{P}^2$. Two discs are removed from each of these surfaces and the resulting boundary circles glued together. What is the surface?

▷ **Exercise 4.16.** Classify the following surfaces:

(1) ABD, BLJ, BJK, ACI, BCH, AGI, EGH, EFH, EJK, LEK, BDE, ACD, CDF, BCF, LBE, EFJ, BFK, ABG, CHI, FHI, LFK, LFJ, DFI, DGI, DEG, BGH.

(2) ABH, AHG, AGO, AOF, AFX, DAX, ADJ, ACJ, ACS, AES, AEZ, ABZ, CSW, SWR, SNR, SEN, EYZ, ZYV, VUY, PQU, ZVW, RVW, BZW, BCW, BCI, BHI, HIL, KHL, GHK, FGK, FGT, POT, OGT, OFK, PKO, PKL, PLQ, MLQ, MRQ, QRV, QUV, FTX, UTX, PTU, UYX, XDY, DEY, DEN, DNJ, MNJ, MNR, MIJ, CIJ, MIL.

(3) ABD, ABE, EBG, EFG, DEF, BDF, FIG, BIG, FIJ, JIN, MJN, MKN, MJF, LFM, LFN, KLN, NFH, NOH, OHL, KLO, LHI, LIM, KMO, IOM, NIO, BCI, ACE, CEI, DEI, DHI, ACD, CDH, BCH, BFH.

▷ **Exercise 4.17.** Identify the surfaces which have planar diagram with the outer edges labeled as below:

(1) $abcdec^{-1}da^{-1}b^{-1}e^{-1}$
(2) $ae^{-1}a^{-1}bdb^{-1}ced^{-1}c^{-1}$
(3) $abc^{-1}d^{-1}ef^{-1}fe^{-1}dcb^{-1}a^{-1}$
(4) $abcdbeafgd^{-1}g^{-1}hcife^{-1}ih^{-1}$

▷ **Exercise 4.18.** If one removes two discs from a surface S and glues the boundary circles of these two discs together, is the result a surface? If so, which surface?

4.5 Surfaces with boundary

A triangulated surface with boundary is a topological space obtained from a set of triangles with edges and vertices identified as for the surfaces in Section 4.1, except some edges will not be identified. These unmatched edges form the boundary of the surface. Surfaces with boundary can be classified by an extension of the method of Theorem 4.14. First, we modify Definition 4.11 to provide for boundary triangles:

(4.16) Definition. *A triangulated surface with boundary is a space with a* *simplicial 2-complex such that:*

(1) *Each edge is identified to at most one other edge.*

(2) *The triangles meeting at a vertex can be labeled $T_1, T_2, \ldots, T_n$ with adjacent triangles in this sequence identified along an edge and T_n either glued to T_1 along an edge, or T_n and T_1 each have one edge on the boundary.*

(3) *No edge not on the boundary can have both vertices on the boundary.*

Condition (3) of Definition 4.16 is added to forbid cases as shown in Figure 4.59.

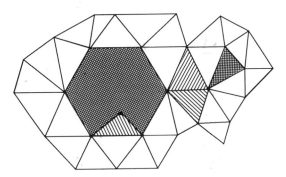

Fig. 4.59. These triangles violate Definition 4.16(3)

This condition is added so that it is possible to clearly identify the vertices and edges of the boundary. If vertices A and B are on the boundary, then the edge AB must also lie on the boundary.

In order to satisfy condition (3), we may have to alter the simplicial complex by dividing some of the triangles and edges into smaller pieces. Taking the barycentric subdivision provides a new triangulation which will then satisfy all the conditions of Definition 4.16. In practice, however, it is more efficient to only subdivide edges and triangles as necessary, thus minimizing the number of new triangles.

(4.17) Theorem. *A compact connected surface with boundary is topologically equivalent to a sphere, the connected sum of n tori, or the connected sum of n projective planes, with a finite number of discs removed.*

Proof. Given a simplicial complex for the surface, modify it as described above so that condition (3) of Definition 4.16 is satisfied. Construct a planar model by first finding all of the boundary edges and vertices from the list of triangles in the complex (these edges will be unmatched); see Figure 4.60. Glue all triangles possessing boundary edges or vertices together to form a planar diagram of the portion of the surface surrounding the boundary. If the boundary has more than one component you will need to do this for each boundary component. You obtain something like Figure 4.60 for each piece of boundary, and some spare triangles.

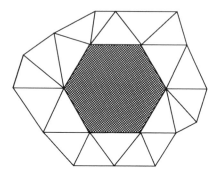

Fig. 4.60. Building the planar model around the boundary

Glue all of these boundary planar diagrams and the extra triangles together to form a planar diagram of the surface which will then have all its holes in the middle (Fig. 4.61).

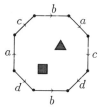

Fig. 4.61. A planar model of a surface with boundary

The outer edges must then all be identified in pairs. Carry out the steps of Theorem 4.14 on this diagram, being careful not to cut through any of the boundary components. The holes may be shrunk and slid around to facilitate the process. As in Theorem 4.14, one obtains one of the desired forms, but punctured with holes topologically equivalent to discs.   □

▷ **Exercise 4.19.**   For a compact triangulable surface, show that the boundary must be homeomorphic to a finite number of disjoint circles.

▷ **Exercise 4.20.**   The cylinder and the Möbius band are surfaces with boundary. Describe each as a sphere, a connected sum of n tori, or a connected sum of n projective planes, with a finite number of discs removed.

▷ **Exercise 4.21.**   Classify the following surfaces with boundary. [Hint: Some of these you may be able to recognize without going through the entire process of Theorem 4.17.]

(1) ABG, BGH, BCH, CDH, DHI, DEI, GHI, AGI, EFI.

(2) ADH, ADF, AFG, CFG, CFI, CIJ, BIJ, BEJ, BEH, DEH.

(3) AEP, ACP, POC, POG, EPG, BTE, EIT, GIM, BNM, CNM, COM, GOM, FDQ, FQR, RQG, SEQ, DQS, CAJ, ADJ, DJK, HJK, CHJ, LES, SLD, IKL, IEL, DKL, HIK, CHF, NFC, BFN, ABE, BMT, MIT, HRG, HFR, BDF, EQG.

(4) DCH, BCH, BGH, GFJ, EJC, BCJ, BIJ, ABI, AFI, FIH, EFH, ECD, EFJ, EDH, BFG.

Chapter 5

The euler characteristic

"The study of mathematics is, if an unprofitable, a perfectly innocent and harmless occupation."

G.H. Hardy

5.1 Topological invariants

In Chapter 4 we proved that all surfaces can be classified into three types: the sphere, connected sums of tori, and connected sums of projective planes. We claimed, but did not really prove, the seemingly obvious fact that these are really different types. As you may have noticed, the obviousness of a fact never stops a mathematician from proving it. Thus, of course, our next mission is to prove that these are truly different shapes. Theoretically, there might be some way of continuously deforming the sphere so that it suddenly acquires a handle, or some way of distorting a torus so that a Möbius band appears inside of it. In the language of Theorem 4.14, there might be a way of cutting and pasting the planar diagram for the Klein bottle so that it becomes the planar diagram for the torus. Lest you lose all faith, I should reassure you that none of these eventualities ever occur, but we never claimed to have investigated all possible cuttings and pastings. Two figures are topologically equivalent (Definition 2.19) if there is a homeomorphism between them. Are we then to consider all possible homeomorphisms of a space to see if one of them could possibly give the second space?

Even without these quibbles, there remains the fact that we do not really have an efficient way of deciding which manifold is represented by pictures such as that in Figure 5.1. Of course, this surface can be classified by either manipulating it in your imagination until it is identifiable, or by cutting it open to form a complex, triangulating the complex, and then going through the process of Theorem 4.14. The first method of attack seems, at best, unreliable, and the second quite tedious.

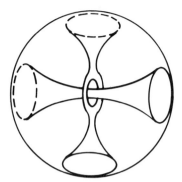

Fig. 5.1. What is this surface?

▷ **Exercise 5.1.**   Classify the surface of Figure 5.1.

It would be preferable to have some property or quantity that could indisputably differentiate between two surfaces. This, however, requires more sophisticated mathematics. In this chapter, we introduce the simplest topological invariant: the euler characteristic. This is the first step in the algebraization of topology: finding something computable to describe the shape of a space.

(5.1) Definition. *The quantity α is a topological invariant if whenever X and Y are topologically equivalent, then $\alpha(X) = \alpha(Y)$.*

Ideally, one would like a topological invariant that would also satisfy the condition that if $\alpha(X) = \alpha(Y)$, then X and Y are homeomorphic. The euler characteristic is too simple to be completely successful at this, though still quite useful. The next chapter is devoted to an introduction of homology theory, which, though more difficult to compute, not only contains all the information carried by the euler characteristic but also more explicit information about the shape of a space.

5.2 Graphs and trees

Before we try to tackle the surfaces again, let us consider simple 1-complexes.

(5.2) Definition. *A graph, Γ, is a connected 1-complex.*

Graphs are built from edges and vertices only. Any two vertices on a graph are connected by some path through the edges and vertices of the

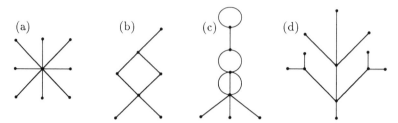

Fig. 5.2. Examples of graphs

graph. Note that some of the graphs in Figure 5.2 have *loops* or *cycles*: paths on the graph that begin and end at the same vertex. These can be considered as redundant paths connecting two vertices.

(5.3) Definition. *A tree, T, is a graph with no cycles.*

Thus, examples (a) and (d) of Figure 5.2 are trees. The other figures have cycles.

▷ **Exercise 5.2.** Below are the numbers of topologically different trees for each number of vertices. Draw these trees.

Table 5.1. Numbers of vertices and trees

number of vertices	1	2	3	4	5	6	7	8
number of trees	1	1	1	2	3	6	11	23

We would like to find a topological invariant that would distinguish the trees among the graphs and tell whether two graphs are topologically equivalent. Count the number of edges and vertices in the examples above, letting v denote the number of vertices and e the number of edges. Note that for the tree of Figure 5.2(a), $v = 7$ and $e = 6$. Figure 5.2(d), another tree, has $v = 12$ and $e = 11$. The actual numbers of edges and vertices do not much to do with anything, but for both of the trees we have one more vertex than edges. Could this be coincidence? Intuitively, a tree is built by starting with a line segment with one edge and two vertices, so $v - e = 1$, and adding branches at one of the already existing vertices of the tree.

Adding a branch in this manner adds one edge and one vertex. Thus, $v - e$ should still be 1. This argument does not use the language of Definition 5.3, so this does not pretend to be a proof. First, give the quantity $v - e$ a name:

(5.4) Definition. *For any graph* Γ, $\chi(\Gamma) = v - e$ *is the euler characteristic of* Γ.

(5.5) Theorem. *Let T be a tree. Then $\chi(T) = 1$.*

Proof. The proof proceeds by induction on e, the number of edges in T. Note that if $e = 0$, then the tree T consists of a single vertex, so $v = 1$. In this case, $\chi(T) = v - e = 1 - 0 = 1$.

Assume that the theorem is true for all trees with less than n edges, and let T be a tree with $e = n$. Choose an edge a of the tree at random, as in Figure 5.3. Designate the endpoints of edge a by P and Q. Note that if edge a is deleted from T, then the tree falls apart into two disconnected subtrees, denoted T_1 and T_2, as in Figure 5.4.

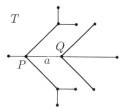

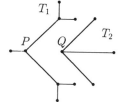

Fig. 5.3. The tree T with edge a **Fig. 5.4.** T with edge a deleted

This must be true, since if $T - \{a\}$ were not disconnected, then there would have to be some other path on $T - \{a\}$ connecting P to Q. But this path joined with edge a would form a cycle from P to Q and back again. This contradicts the assumption that T is a tree with no cycles. The subtrees T_1 and T_2 have fewer than n edges, so by the induction hypothesis,

$$\chi(T_1) = v_1 - e_1 = 1$$
$$\chi(T_2) = v_2 - e_2 = 1$$

Note that every vertex of T is in either T_1 or in T_2, but not in both, so $v = v_1 + v_2$. The edges of T consist of the edge a, the edges of T_1, and the edges of T_2, so $e = 1 + e_1 + e_2$. Thus,

$$\chi(T) = v - e$$
$$= (v_1 + v_2) - (1 + e_1 + e_2)$$
$$= (v_1 - e_1) + (v_2 - e_2) - 1$$
$$= 1 + 1 - 1 = 1$$

$\square$

All trees have the same euler characteristic. This, of course, does not mean that all trees are topologically equivalent, but it does tell something about the shape.

Returning to the examples of Figure 5.2 which are not trees, note that (b) has seven vertices and seven edges and, thus, has $\chi = 0$. Figure 5.2(c) has seven vertices and ten edges, so $\chi = -3$. The euler characteristic seems to be telling us that these are not trees, but these numbers have further significance. Note that Figure 5.2(b) has one loop.

The loops of Figure 5.2(c) are harder to count. Label the edges of the figure as in Figure 5.5. The edge j forms one loop, and the edges g and h make another. The bit comprising edges d, e, and f is trickier: Consider one loop formed by edges d and e, and another by e and f, but do not count the circuit formed by d and f as a separate loop. In justification, we might say that the d–f loop is just the d–e loop combined with the e–f loop. Another argument is that there must be one path from P to Q on any connected graph, but here we have three; two are redundant, so we count this as two loops. Therefore, we consider Γ to have four distinct cycles.

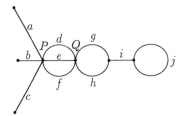

Fig. 5.5. The graph Γ

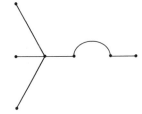

Fig. 5.6. A spanning tree for Γ

Another way to count loops is to remove edges, as few as possible, from the redundant loops in Γ, until a tree is left. This is called a *spanning tree* of Γ, and is not unique. Note that the spanning tree of Γ will be connected and will contain all the vertices of Γ. The number of edges removed from Γ to make the spanning tree will equal the number of loops in Γ. (See Fig. 5.6.)

The simplest graphs are trees and have $\chi = 1$. Our example from Figure 5.2(b) has one loop and $\chi = 0$. The example from Figure 5.2(c) has four loops and $\chi = -3$. These more complicated graphs can be constructed from a tree by adding loops or redundant edges. The numbers that we have calculated seem to imply that each loop reduces the euler characteristic by one.

(5.6) Theorem. *Let Γ be a graph with n distinct cycles. Then $\chi(\Gamma) = 1 - n$.*

Proof. For each of the n cycles in Γ, we can remove one edge without disconnecting the graph, since each loop implies a redundancy of paths between two vertices. After doing this, we have a graph with no cycles, i.e., a spanning tree T, and by Theorem 5.5, $\chi(T) = 1$. The tree T is not unique, but it does contain all the vertices of Γ, with n fewer edges.

$$1 = \chi(T) = v - (e - n) = (v - e) + n = \chi(\Gamma) + n$$

Therefore, $\chi(\Gamma) = 1 - n$.   □

The euler characteristic does not give a precise description of a graph but does give valuable information about the shape. Since the euler characteristic seems useful, we will now prove that it is a topological invariant.

(5.7) Theorem. *Let Γ_1 and Γ_2 be topologically equivalent graphs. Then $\chi(\Gamma_1) = \chi(\Gamma_2)$.*

Proof. The hypothesis is that $|\Gamma_1|$ and $|\Gamma_2|$ are homeomorphic. A vertex with only one edge coming out of it is called an *endpoint vertex*; see Figure 5.7. A vertex where more than two branches meet (and we let an edge which begins and ends at the same vertex such as the edge j in Figure 5.5(c) count as two branches) is called a *branch vertex*; see Figure 5.8.

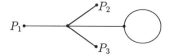

Fig. 5.7. P_1, P_2, P_3 are endpoint vertices

Fig. 5.8. Q_1, Q_2 are branch vertices

It is clear that a homeomorphism must take endpoints to endpoints and branch vertices to vertices with the same number of branches. Thus, the endpoint vertices and the branch vertices of Γ_1 and Γ_2 coincide, though their other vertices may not. Let Γ be the graph with the same shape as Γ_1 and Γ_2, but with all the vertices of both, as in Figure 5.9. The graph Γ is called a *common subdivision* of Γ_1 and Γ_2 since it is obtained by subdividing the edges of either. Both $|\Gamma_1|$ and $|\Gamma_2|$ are topologically equivalent to $|\Gamma|$, and Γ can be derived by adding vertices in the middle of the edges of either Γ_1 or Γ_2.

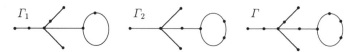

Fig. 5.9. The graphs Γ_1, Γ_2, and Γ

Note that adding a vertex in the middle of an edge does not change the euler characteristic, as illustrated in Figures 5.10 and 5.11.

Fig. 5.10. $v = 2$, $e = 1$, so $\chi = 1$ **Fig. 5.11.** $v = 3$, $e = 2$, so $\chi = 1$

Thus, both Γ_1 and Γ_2 have the same euler characteristic as Γ, so $\chi(\Gamma_1) = \chi(\Gamma_2)$.   □

▷ **Exercise 5.3.**   Draw graphs with:

(1) only one endpoint
(2) no endpoints
(3) all vertices are endpoints
(4) no branch points
(5) all vertices are branch points

(5.8) Definition. *A complete graph, K_n, on n vertices is one where every pair of vertices is connected by one edge. (See Figures 5.12 and 5.13.)*

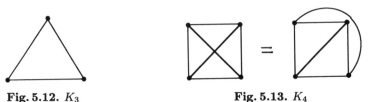

Fig. 5.12. K_3 **Fig. 5.13.** K_4

▷ **Exercise 5.4.**

   (1) Draw K_2, K_5, K_6.
   (2) Deduce a rule for the number of edges in K_n.

5.3 The euler characteristic and the sphere

We now generalize the euler characteristic to complexes of higher dimensions. We first study how much information can be gotten from the euler characteristic, and then, once we have convinced ourselves that it is worth keeping, show that it is a topological invariant. In view of Theorems 5.5 and 5.6, it is clearly too much to ask that the euler characteristic tell us exactly what a space looks like, but it may well give information about similarities and differences among spaces, as it did for graphs and trees.

All spaces are represented by finite complexes, as in Chapter 4. Let $\#(S)$ denote the number of elements in a finite set S.

(5.9) Definition. *Let K be a complex. The euler characteristic of K is*

$$\chi(K) = \#(\text{0-cells}) \; - \#(\text{1-cells}) + \#(\text{2-cells}) - \#(\text{3-cells.}) \pm \cdots$$

For 2-complexes, let $f = \#\{\text{faces}\}$, $e = \#\{\text{edges}\}$, and $v = \#\{\text{vertices}\}$, and then the euler characteristic may be written as

$$\chi(K) = v - e + f$$

Consider a polygon with n sides, illustrated in Figure 5.14. The complex K has one face, n edges, and n vertices, so $\chi(K) = 1 - n + n = 1$.

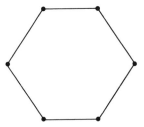

Fig. 5.14. For this complex K on the disc, $\chi(K) = 1$

Another example is the complex K' given by the standard planar diagram of the sphere $\mathbb{S}^2$ in Figure 5.15. This complex K' has one face, one edge (a), and two vertices (P and Q), so $\chi(K') = 1 - 1 + 2 = 2$. However, there are many different complexes which could also represent this sphere.

It is natural to ask if these also give $\chi = 2$, and this is the intent of the next theorem.

Fig. 5.15. For this complex K' on the sphere, $\chi(K') = 2$

(5.10) Theorem. *Any 2-complex, K, such that $|K|$ is topologically equivalent to the sphere, has euler characteristic $\chi(K) = 2$.*

Proof. Each face of the 2-complex is a polygon. Any polygon can be divided into triangles by introducing a new vertex in the interior of the polygon and connecting this vertex to each of the vertices of the polygon, as in Figure 4.35. If the polygon originally had n edges and so n vertices, then $\chi = 1 - n + n = 1$. The new subdivision has n faces, $2n$ edges, and $n + 1$ vertices, so $\chi = n - 2n + (n + 1) = 1$. Thus, there is no net change in the euler characteristic. Do this for each (non-triangular) face of K to obtain a new complex K' with $|K'| = |K|$ and $\chi(K') = \chi(K)$ and K' has only triangles for faces. Note that we are not claiming that K' is a triangulation (recall Definition 4.9) but that K' has only triangular faces. Thus, K' looks like a sphere covered by triangles, as in Figure 5.16. We must show that $\chi(K') = 2$. Let $\chi = \chi(K')$.

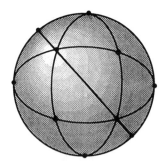

Fig. 5.16. K' is a complex on the sphere with only triangular faces

Step 1: Remove the interior of one of the triangles. We then have a sphere with one hole, and a new complex with one less face, but the same edges and vertices so that the new euler characteristic is $\chi - 1$. Note that the new figure is still a 2-complex. A sphere with a hole can be flattened into a disc, by the stereographic projection of Definition 2.21. This projection gives a complex K'' on a subset of $\mathbb{R}^2$ homeomorphic to the disc. Note that K' and K'' have the same number of triangles, edges, and vertices, so the euler characteristic of the complex K'' is $\chi - 1$.

Step 2: Remove the triangles of K'' one by one, starting with those bordering the hole. Each triangle must be removed in such a manner that the topological space remaining after the triangle is deleted is a 2-complex with only triangular faces, so some of the edges and vertices may be left behind. There are three cases:

(1) *Case 1: The triangle has one edge along the boundary (Fig. 5.17).*

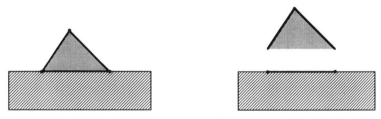

Fig. 5.17. In this case we removed one face, two edges, and one vertex, so the new euler characteristic is $(\chi - 1) - (1 - 2 + 1) = \chi - 1$

(2) *Case 2: The triangle has two edges along the boundary (Fig 5.18).*

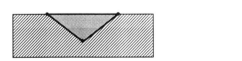

Fig. 5.18. In this case we removed one face, one edge, and no vertices, so the new euler characteristic is $(\chi - 1) - (1 - 1 + 0) = \chi - 1$

(3) *Case 3: The triangle has one vertex along the boundary (Fig 5.19).*

Fig. 5.19. In this case we removed one face, three edges, and two vertices, so the new euler characteristic is $(\chi - 1) - (1 - 3 + 2) = \chi - 1$

Step 3: Note that none of the cases of Step 2 result in any change in the euler characteristic, so we still have $\chi - 1$. Eventually we are left with one triangle which must have euler characteristic $\chi - 1$. A triangle has $f = 1$, $e = 3, v = 3$, so $\chi - 1 = (1 - 3 + 3) = 1$. Thus, the original euler characteristic was $\chi(K) = 2$. □

For an extremely detailed and thought-provoking analysis of this proof as an example of the techniques of mathematical thought, see Imre Lakatos' *Proofs and Refutations*. Theorem 5.10 shows that the euler characteristic is a topological invariant for the sphere.

▷ **Exercise 5.5.** Explain why the following examples (Figs. 5.20 and 5.21) are not counterexamples to Theorem 5.10:

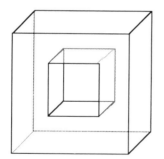

Fig. 5.20. K_1, the picture frame, considered as a polyhedron with 10 sides

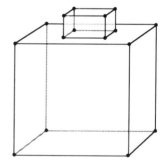

Fig. 5.21. K_2, the crested cube, considered as a polyhedron with 11 sides

The converse of Theorem 5.10 is not true, as there are complexes with $\chi = 2$ which are not homeomorphic to the sphere; see Figure 5.22.

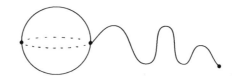

Fig. 5.22. Some complexes with $\chi = 2$, which are not homeomorphic to the sphere

There are, of course, many complexes representing the sphere, but some have been of especially long-standing interest:

(5.11) Definition. *A regular polyhedron is a polyhedron whose faces all have the same number of sides, and which also has the same number of faces meeting at each vertex.*

The regular polyhedra which are topologically equivalent to the sphere are known as the *Platonic solids* (Fig. 5.23) and have been studied for over 2000 years. The five Platonic solids are:

(1) the tetrahedron: each face has three sides, three faces meet at each vertex
(2) the cube: each face has four sides, three faces at each vertex
(3) the octahedron: each face has three sides, four faces at each vertex
(4) the icosahedron: each face has three sides, five faces at each vertex
(5) the dodecahedron: each face has five sides, three faces at each vertex.

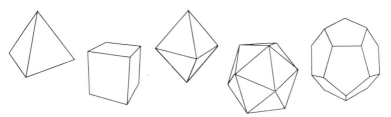

Fig. 5.23. Tetrahedron, cube, octahedron, icosahedron, dodecahedron

(5.12) Theorem. *The Platonic solids are the only regular polyhedra topologically equivalent to a sphere.*

Proof. Given a regular polyhedron K which is topologically equivalent to the sphere, let f denote the number of faces in K, e the number of edges, and v the number of vertices. Let n be the number of sides on each face, and let m be the number of faces meeting at each vertex. We know from Theorem 5.10 that $\chi(K) = f - e + v = 2$. Consider the pile of polygons before they are glued together to form the polyhedron K, and let f' denote the number of faces before assembly, e' the number of edges, and v' the number of vertices. For example, the deconstruction of the tetrahedron is illustrated in Figure 5.24.

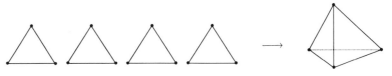

Fig. 5.24. $f' = 4, e' = 12, v' = 12$ and $f = 4, e = 6, v = 4$

The number of polygons or faces is the same before or after assembly, so $f = f'$. Before gluing, each face has n edges and n vertices, so $nf = e' = v'$. The edges are glued together in pairs in K, so $e' = nf = 2e$. In assembling K, m faces meet at each vertex of K, so m vertices from m unglued faces are glued together to make one vertex in K, and $v' = mv$. Thus, $v' = mv = nf = 2e$. Therefore,

$$2 = v - e + f$$
$$= \frac{2e}{m} - e + \frac{2e}{n}$$
$$= e\left(\frac{2}{m} - 1 + \frac{2}{n}\right)$$
$$\frac{1}{e} = \frac{1}{m} - \frac{1}{2} + \frac{1}{n}$$

Note that e, n, m must, of course, be integers, and that $e > 2$, $n > 2$, $m > 2$, so $\frac{1}{e} < \frac{1}{2}$. Since diophantine equations (equations with only integer solutions allowed) such as the one above are rather difficult to solve, we will analyze each possible case separately:

Case 1: $n = 3$ (the polygons are triangles)

$$\frac{1}{2} > \frac{1}{e} = \frac{1}{m} - \frac{1}{2} + \frac{1}{3} = \frac{1}{m} - \frac{1}{6} > 0$$
$$\frac{2}{3} > \frac{1}{m} > \frac{1}{6}$$

$$\frac{3}{2} < m < 6$$

Since $m > 2$, the only possibilities are $m = 3$, 4, or 5.

(1) If $m = 3$, then $\frac{1}{e} = \frac{1}{3} - \frac{1}{2} + \frac{1}{3} = \frac{1}{6}$ so $e = 6$, $f = \frac{2e}{n} = 4$, and $v = \frac{2e}{m} = 4$. This is the tetrahedron.

(2) If $m = 4$, then $\frac{1}{e} = \frac{1}{4} - \frac{1}{2} + \frac{1}{3} = \frac{1}{12}$, so $e = 12$, $f = 8$, and $v = 6$. This is the octahedron.

(3) If $m = 5$, then $\frac{1}{e} = \frac{1}{5} - \frac{1}{2} + \frac{1}{3} = \frac{1}{30}$, so $e = 30$, $f = 20$, and $v = 12$. This is the icosahedron.

Case 2: $n = 4$ (faces are squares)

$$\frac{1}{2} > \frac{1}{e} = \frac{1}{m} - \frac{1}{2} + \frac{1}{4} = \frac{1}{m} - \frac{1}{4} > 0$$

$$\frac{3}{4} > \frac{1}{m} > \frac{1}{4}$$

$$\frac{4}{3} < m < 4$$

Since m must be greater than 2, the only possibility is $m = 3$, and then $\frac{1}{e} = \frac{1}{3} - \frac{1}{2} + \frac{1}{4} = \frac{1}{12}$, so $e = 12$, $f = 6$, $v = 8$. This is the cube.

Case 3: $n = 5$ (faces are pentagons)

$$\frac{1}{2} > \frac{1}{e} = \frac{1}{m} - \frac{1}{2} + \frac{1}{5} = \frac{1}{m} - \frac{3}{10} > 0$$

$$\frac{4}{5} > \frac{1}{m} > \frac{3}{10}$$

$$\frac{5}{4} < m < \frac{10}{3}$$

The only possibility is $m = 3$, and then $e = 30$, $f = 12$, $v = 20$. This is the dodecahedron.

Case 4: $n \geq 6$ (faces are hexagons or bigger)

$$0 < \frac{1}{e} = \frac{1}{m} - \frac{1}{2} + \frac{1}{n} \leq \frac{1}{m} - \frac{1}{2} + \frac{1}{6} = \frac{1}{m} - \frac{1}{3}$$

$$\frac{1}{m} > \frac{1}{3}$$

$$m < 3$$

This cannot happen, so there are only five solutions. $\square$

▷ **Exercise 5.6.** Why must $n \geq 3$ and $m \geq 3$?

It should be noted that there are an infinitude of irregular polyhedra, as well as regular polyhedra which are not equivalent to spheres.

5.4 The euler characteristic and surfaces

The euler characteristic is strikingly easy to compute, and it really seems odd that it should be a topological invariant. After all, it is computed from a cell decomposition, and it is very easy to come up with a lot of different complexes on a single surface, all with varying numbers of faces, edges, and vertices. But somehow by taking the alternating sum, one arrives at a quantity which depends only on the underlying shape and not on the particular complex. This was proven for the sphere in Theorem 5.10, and we wish to show that the euler characteristic is a topological invariant for all surfaces.

▷ **Exercise 5.7.**   Compute the euler characteristics for the torus, projective plane, Klein bottle, cylinder, and Möbius band. You may use any complexes you like to represent these surfaces.

The technique of Theorem 5.10 does not generalize to other surfaces, so we must use another method. In the proof of Theorem 5.7, we showed that two topologically equivalent graphs had the same euler characteristic by finding a common subdivision. This provides us with a clue to pursue.

We may assume that we have two cell complexes K and K' such that $|K|$ and $|K'|$ are topologically equivalent surfaces, and we wish to show that $\chi(K) = \chi(K')$. We may also apply the trick used in the first paragraph of the proof of Theorem 5.10 to alter K and K' so that they only have triangular cells, without changing the euler characteristic.

▷ **Exercise 5.8.**   Let T be a triangle, considered as a complex. If T' is the barycentric subdivision of T, show that $\chi(T) = \chi(T')$.

Apply Exercise 5.8 to each of the triangles of the complexes K and K' (twice if necessary) to obtain simplicial complexes which we will still denote K and K', with no change in the euler characteristic. Now we have two simplicial complexes K and K' such that $|K|$ and $|K'|$ are topologically equivalent surfaces. By Theorem 4.14 (or Theorem 4.17 if $|K|$ and $|K'|$ are surfaces with boundary), we know that K and K' must be two different complexes representing the same surface and that surface must be either the sphere, a connected sum of tori, or a connected sum of projective planes. Theorem 4.14 (or 4.17) gives a standard method for reducing the triangulations K and K' to one of the standard planar diagrams, representing the standard complexes for the surfaces. If we can show that the seven-step process of Theorem 4.14 does not alter the euler characteristic, then it would follow that

$$\chi(K) = \chi(\text{standard planar diagram for the surface})$$
$$\chi(K') = \chi(\text{standard planar diagram for the same surface})$$

and so $\chi(K) = \chi(K')$.

(5.13) Theorem. *The euler characteristic is a topological invariant for compact connected surfaces, and so does not depend on the representation.*

Proof. As outlined in the paragraph above, we may assume that K and K' are two simplicial complexes such that $|K|$ and $|K'|$ are topologically equivalent surfaces and show that the process of triangulating the complexes and the seven-step process of Theorem 4.14 does not alter the euler characteristic. Both of these procedures involve only three types of maneuvers, of the types shown in Figures 5.25 – 5.27. Let K be the original complex and let complex L be obtained by a single operation of one of these three types from K. Denote the faces, edges, and vertices of K by f, e, v, and those of L by f', e', v'.

Type 1: Add an edge between two vertices of a polygon.

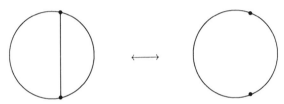

Fig. 5.25. Type 1 operation

Note that $f' = f+1$, $e' = e+1$, $v' = v$, so $\chi(L) = (f+1)-(e+1)+v = \chi(K)$. This move or its inverse is used in the barycentric subdivision and when cutting along a new edge or gluing along an edge, as in Steps 4,5, and 6 of Theorem 4.14 or in Lemma 4.15.

Type 2: Add a vertex in the interior of a polygon and an edge from this vertex to a boundary vertex of the polygon.

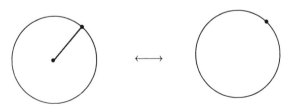

Fig. 5.26. Type 2 operation

Note that $f' = f$, $e' = e + 1$, and $v' = v + 1$, so

$$\chi(L) = f' - e' + v' = f - (e + 1) + (v + 1) = \chi(K)$$

This maneuver or its inverse is used in the barycentric subdivision and in Step 1 and Step 3 of Theorem 4.14.

Type 3: Add a vertex in the interior of an edge.

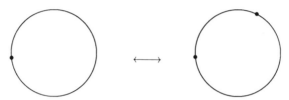

Fig. 5.27. Type 3 operation

Note that $f' = f, e' = e+1, v' = v+1$, so $\chi(L) = f-(e+1)+(v+1) = \chi(K)$. This move is used in the barycentric subdivision and in Step 2 of Theorem 4.14, when a string of edges is relabeled as a single edge.

The process of Theorem 4.14 involves a finite number of these three operations, and so does not affect the euler characteristic. The euler characteristic depends only on the type of surface and not on the triangulation.

$\square$

$\triangleright$ **Exercise 5.9.** Break down the process of taking the barycentric subdivision of a triangle into a sequence of moves of the types of Theorem 5.13.

The method of proof used in Theorem 5.13 will not work for complexes in general nor for higher-dimensional manifolds, since there is no known classification theorem like Theorem 4.14.

We may now compute the euler characteristic for all surfaces from the standard planar diagrams in Chapter 4; see Table 5.2.

Table 5.2. Euler characteristics

surface	f	e	v	χ
sphere	1	1	2	2
n tori	1	$2n$	1	$2 - 2n$
m projective planes	1	m	1	$2 - m$

Note that the orientable surfaces, the sphere and tori, have even euler characteristic, whereas the non-orientable surfaces, projective planes, can have either even or odd euler numbers. The euler characteristic can considerably shorten the process of classifying surfaces.

(5.14) Theorem. *S_1, S_2 are compact, connected surfaces without boundary. Then S_1 is topologically equivalent to S_2 if and only if $\chi(S_1) = \chi(S_2)$ and either both are orientable or both are non-orientable.*

Proof. This follows immediately from Theorems 4.14 and 5.13.   □

In practice, the most efficient way of identifying a surface from a triangulation is first to use Definition 4.11 to check the triangulation to determine that it is indeed a surface and what, if any, the boundary may be, then compute the euler characteristic. This is most easily done from the planar diagram constructed in Step 1 of Theorem 4.14. For surfaces without boundary, if $\chi = 2$, the surface must be a sphere; if χ is odd, the surface is a sum of projective planes; and only if χ is even need one go through the entire process of Theorem 4.14 in order to determine the surface. Even then, one can stop after Step 4 is completed (if a twisted pair is present at this point, the surface is non-orientable and so is a sum of projective planes).

▷ **Exercise 5.10.**   Reclassify the surfaces of Exercises 4.16 and 4.17 as efficiently as you can.

▷ **Exercise 5.11.**

(1) Prove $\chi(S_1 \# S_2) = \chi(S_1) + \chi(S_2) - 2$ for any surfaces S_1 and S_2.
(2) Use part (1) to calculate the euler characteristic of the n-handled torus and the connected sum of n projective planes.

▷ **Exercise 5.12.**   If K_1 and K_2 are simplicial complexes such that $K_1 \cap K_2$ is also a simplicial complex, prove that

$$\chi(K_1 \cup K_2) = \chi(K_1) + \chi(K_2) - \chi(K_1 \cap K_2)$$

The genus of a surface is a related invariant which counts the number of tori or handles in a surface for the orientable case, and the number of twisted pairs or projective planes for the non-orientable case.

(5.15) Definition. *Let S be a compact surface. The genus of S is*

$$g(S) = \begin{cases} \frac{1}{2}(2 - \chi) & \text{if } S \text{ is orientable} \\ 2 - \chi & \text{if } S \text{ is non-orientable} \end{cases}$$

Thus, the Klein bottle is referred to as the non-orientable surface with genus 2, and the n-handled torus as the orientable surface with genus n. The genus provides no additional information but is a way of converting the euler characteristic into a number which is directly related to the physical properties of the surface.

To determine whether two surfaces with boundary are topologically equivalent, we must check the number of holes as well as the orientability and the euler characteristic.

(5.16) Definition. *If S is a surface with boundary, the associated surface (without boundary) is S^*, where S^* is S with a disc sewn onto each of the boundary circles.*

In other words, S^* is the surface S with all its holes patched. By Exercise 4.19, the boundary of S consists of disjoint circles. If S has k holes or boundary components, then S^* can be considered as S with k discs or 2-cells added so

$$\chi(S^*) = \chi(S) + k$$

(5.17) Theorem. *Let S_1 and S_2 be compact connected surfaces with boundary. Then S_1 is topologically equivalent to S_2 if and only if they have the same number of boundary components, both are orientable or both non-orientable, and they have the same euler characteristic.*

Proof. If S_1 and S_2 are topologically equivalent, then their boundaries are also homeomorphic, so they have the same number of boundary components. If one contains a Möbius band, then the other also does, and they must have the same orientability type. Since they will thus have the same standard planar diagram, they have the same euler characteristic.

Conversely, if S_1 and S_2 have the same number of holes, orientability type, and $\chi(S_1) = \chi(S_2)$, consider the associated surfaces S_1^* and S_2^*. For S_1 and S_2, let $k = \#$ (holes in S_1) $= \#$ (holes in S_2). Since $\chi(S_1) = \chi(S_2)$,

$$\chi(S_1^*) = \chi(S_1) + k$$
$$= \chi(S_2) + k$$
$$= \chi(S_2^*)$$

If S_1 and S_2 have the same orientability type, then so do S_1^* and S_2^*. By Theorem 5.14, S_1^* is topologically equivalent to S_2^*. By removing k discs from each of S_1^* and S_2^*, S_1 is topologically equivalent to S_2.   □

(5.18) Definition. *The genus of a compact surface with boundary S is $g(S) = g(S^*)$ where S^* is the associated surface without boundary.*

▷ **Exercise 5.13.** Find a formula for the genus of a compact connected surface with boundary in terms of the euler characteristic and k, the number of boundary components.

▷ **Exercise 5.14.** Reclassify the surfaces of Exercise 4.21 as efficiently as you can.

▷ **Exercise 5.15.** For any triangulation of a compact surface (without boundary), let t = number of triangles, e = number of edges, and v = number of vertices.

  (1) Prove that $3t = 2e$, $e = 3(v - \chi)$ and $v \geq \frac{1}{2}(7 + \sqrt{49 - 24\chi})$ [Hint: see Exercise 5.4.]

  (2) Find the minimal values of t, e, and v for $\mathbb{S}^2$, $\mathbb{P}^2$, and $\mathbb{T}^2$.

  (3) Find the corresponding minimal triangulations for these surfaces.

▷ **Exercise 5.16.** The Klein bottle, $\mathbb{K}^2$, is the sole exception to the existence of minimal triangulations for surfaces (as in Exercise 5.15 above), since eight, rather than seven, vertices are required. Find such a triangulation. Note that this will not be unique.

▷ **Exercise 5.17.** Let S be a compact connected surface which may or may not have boundary, and $\chi(S) = 0$. List all the possibilities for S.

5.5 Map-coloring problems

In this section we discuss an application of the euler characteristic: map-coloring problems. The object is to find the minimum number of colors needed to color all possible maps (assume you have been condemned to repeat fourth grade on a limited budget). The first thing to note is that the countries can be regarded as polygons, the borders with other countries as edges, and place vertices wherever three or more countries meet. For example, in Figure 5.28 note that Tennessee has eight edges and eight vertices.

Fig. 5.28. Tennessee and its neighbors

Countries that share a border or edge must be colored differently, but one is allowed to color countries which meet only at a vertex the same color, as in a checkerboard. We do not allow non-connected countries.

(5.19) The Four-Color Conjecture. *Four colors are enough to color all possible maps drawn on the plane.*

The conjecture that only four colors are needed to color any map drawn on the plane was made by Francis Guthrie in 1852 while, oddly enough, coloring a map. The first proof was published in 1878 by Kempe, but this contained an error found in 1890 by Heawood, who salvaged enough to prove that five colors are sufficient to color any map drawn on the plane. The conjecture was finally confirmed in 1976 by Appel and Haken, after a century of false proofs and refinements of techniques. There are, of course, maps on the plane that do not require four colors (imagine flooding all the continents and painting the map solid blue), but the problem we are discussing is how many crayons one must have on hand to be sure of being able to color any map at all and be able to make adjoining countries different colors.

The problem of coloring a map on the plane is equivalent to the problem on the sphere. If a map on the sphere requires N colors, one can remove one point from the interior of a country on a map on the sphere and stretch the hole open, using stereographic projection, to obtain a map on the plane. The country which had the point removed becomes a sort of ocean surrounding the other countries, and one has a map on the plane using N colors. It should be noted that this projection wildly distorts the relative sizes of countries.

Conversely, if all possible maps on the plane require at most N colors, then given any particular map, we can reverse the process above to get a map on a punctured sphere. If the puncture occurs in the middle of a country, as can be easily arranged, then we obtain a map on the sphere requiring no more than N colors.

If S is a surface, a map drawn on the surface S is represented by a complex K on S. Let f denote the number of faces or polygons in K, e the number of edges, and v the number of vertices.

(5.20) Lemma. *Let S be a surface, and let N be some positive integer. If $\frac{2e}{f} < N$ for all complexes K on S, then N colors are enough to color all possible maps drawn on S.*

Proof. Since S is a surface, each edge in K is identified to exactly one other edge in building the complex. Thus, $2e$ is the total number of edges in the polygons before assembly, and f is the total number of polygons. The number $\frac{2e}{f}$ can be thought of as the average number of edges per polygon. Choose an integer $N > \frac{2e}{f}$ for all complexes K on S.

The proof proceeds by induction on f. Note that the lemma is obviously true if $f < N$, since if there are fewer polygons than colors, one can paint each country a different color.

If we assume the lemma is true for $f = m$, so any complex with m 2-cells can be colored with N colors, we must show that any complex K with $(m+1)$ faces can also be colored with N colors. Since $\frac{2e}{f} = \frac{2e}{m+1} < N$ for the complex K, the average number of edges per polygon is less than N, so there is a 2-cell in K with fewer than N edges. Designate this particular polygon by P. We can eliminate P by shrinking the polygon to a point by a process called *radial projection* (Fig. 5.29).

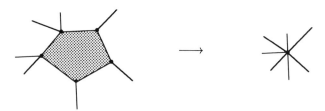

Fig. 5.29. Radial projection

Thus, a new complex K' is formed with one less face than K, so K' has m faces. Since we are assuming the theorem is true for any complex with m faces, K' can be colored with only N colors. The coloring on K' gives a coloring on all of K except for the country P. The face P was chosen because it had fewer than N neighbors, so at least one of the N colors was not used in the countries surrounding P. We can color the face P with this unused color to get a coloring of K using only N colors.   □

The quantity $\frac{2e}{f}$ in the lemma depends on the particular complex or map, rather than the type of surface, but is related to the surface by the euler characteristic.

(5.21) Lemma. *Let S be a surface. Then*

$$\frac{2e}{f} \le 6\left(1 - \frac{\chi(S)}{f}\right)$$

for any complex K on S.

Proof. Consider the pile of polygons which will be glued together to form K. In any polygon, the number of edges is equal to the number of vertices, and the edges of these polygons are matched in pairs in assembling K, so $2e =$ (number of edges before assembling K) = (number of vertices before assembly). After gluing to form K, each vertex in K is formed by gluing at least three polygons together. Thus,

$$3v \le \text{(number of vertices before assembly)} = 2e$$

Using the euler characteristic of S we get

$$\chi(S) = v - e + f$$
$$\chi(S) - f = v - e \le \frac{2e}{3} - e = -\frac{1}{3}e$$

$$3(f - \chi(S)) \geq e$$

Therefore,

$$\frac{2e}{f} \leq \frac{6(f - \chi(S))}{f} = 6\left(1 - \frac{\chi(S)}{f}\right)$$

□

For the torus or Klein bottle, $\chi(S) = 0$ so $\frac{2e}{f} \leq 6$. By Lemma 5.20, if $\frac{2e}{f} < N$, then N colors suffice. Therefore, if we choose $N = 7$, it follows that seven colors are enough to color any map on the torus or the Klein bottle. However, we must ask if all of these colors are necessary. Might we make do with fewer?

(5.22) Theorem. *Seven colors are necessary and sufficient to color all possible maps on the torus.*

Proof. By Lemmata 5.20 and 5.21, seven colors are sufficient. That all seven are necessary is shown by the map of Figure 5.30.

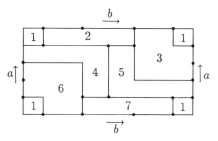

Fig. 5.30. A map on the torus requiring seven colors

After assembly, each country borders all the other countries, so all seven colors are needed.　□

▷ **Exercise 5.18.** Draw a map on the Klein bottle which requires at least six colors. (There are many such.)

For the sphere and projective plane, $\chi(S) > 0$ so $\frac{2e}{f} < 6$. Therefore, no more than six colors are needed to color any map on the sphere (and thus on the plane) or the projective plane. That at least four colors are needed to color all maps in the plane is shown by the example in Figure 5.31. Thus, the minimum number of colors needed for all possible maps on the sphere is either four, five, or six.

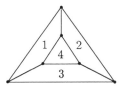

Fig. 5.31. A map on the plane requiring four colors

(5.23) Theorem. *Six colors are necessary and sufficient to color all possible maps on the projective plane.*

Proof. By Lemmata 5.20 and 5.21, six colors are sufficient to color all possible maps on the projective plane. That all six colors are necessary is shown in the following exercise.   □

▷ **Exercise 5.19.** Find a map on the projective plane that requires six colors.

▷ **Exercise 5.20.** Find the minimal number of colors needed to color all maps on a Möbius band.

Heawood conjectured a formula for the number of colors necessary and sufficient to color all possible maps for each surface.

(5.24) Definition. *Let S be a surface. The Heawood number of S is*

$$H(S) = \left\lfloor \frac{7 + \sqrt{49 - 24\chi(S)}}{2} \right\rfloor$$

Note: $\lfloor x \rfloor$ *denotes the greatest integer less than or equal to x.*

Table 5.3 presents the Heawood numbers for the surfaces we have discussed.

Table 5.3. Heawood numbers

	$\mathbb{S}$	$\mathbb{P}$	$\mathbb{T}$	$\mathbb{K}$	$3\mathbb{P}$	$2\mathbb{T}$	$4\mathbb{P}$
χ	2	1	0	0	-1	-2	-2
H	4	6	7	7	7	8	8

(5.25) Theorem. *Let S be a compact connected surface without boundary. Then $H(S)$ colors are sufficient to color any map on S.*

Proof. If $\chi(S) = 2$, then $S = \mathbb{S}^2$ and this is the (very difficult) four-color conjecture, and we will not prove this case. If $\chi(S) = 1$, then S is the projective plane and $H(S) = 6$. This case has already been addressed in Theorem 5.23. If $\chi(S) = 0$, then S is either the torus $\mathbb{T}^2$ or the Klein bottle $\mathbb{K}^2$, and $H(S) = 7$. By the remarks preceding Theorem 5.22, seven colors are sufficient to color $\mathbb{T}^2$ or $\mathbb{K}^2$.

Assume $\chi(S) < 0$. When searching for the minimum number N of colors needed for all possible complexes on a surface S, we may assume that $f > N$ since otherwise we have as many colors as countries.

$$\frac{2e}{f} \leq 6\left(1 - \frac{\chi(S)}{f}\right) \qquad \text{by Lemma 5.21}$$

$$\leq 6\left(1 - \frac{\chi(S)}{N+1}\right) \qquad \text{since } f \geq N+1$$

Then $\frac{2e}{f} < N$ is true if we choose N so that

$$N > 6\left(1 - \frac{\chi(S)}{N+1}\right)$$
$$N(N+1) > 6(N+1) - 6\chi(S)$$

Complete the square:

$$N^2 - 5N + \frac{25}{4} > 6 - 6\chi(S) + \frac{25}{4}$$

$$\left(N - \frac{5}{2}\right)^2 > \frac{49}{4} - 6\chi(S)$$

$$\left|N - \frac{5}{2}\right| > \sqrt{\frac{49 - 24\chi(S)}{4}}$$

$$N > \frac{5}{2} + \frac{\sqrt{49 - 24\chi(S)}}{2}$$

Note that N must be more than $\frac{5+\sqrt{49-24\chi(S)}}{2}$ in order to be certain that $N > \frac{2e}{f}$ and so the hypotheses of Lemma 5.20 are satisfied. We want N to be no larger than necessary, so choose N satisfying

$$\frac{5 + \sqrt{49 - 24\chi(S)}}{2} < N \leq \left\lfloor \frac{5 + \sqrt{49 - 24\chi(S)}}{2} \right\rfloor + 1 = H(S)$$

By Lemma 5.20, $H(S)$ colors suffice to color all possible maps drawn on S. $\qquad\square$

It should be emphasized that the theorem says that $H(S)$ colors are enough to color the map. Whether you really need all of them is another question.

(5.26) Heawood's Conjecture. *Let S be a compact connected surface without boundary. Then $H(S)$ colors are necessary and sufficient to color any map on S.*

Heawood (1890) found the example of Theorem 5.22, thus proving the case of the torus. The case of the projective plane is Theorem 5.23. Ringel and Youngs proved the necessity of $H(S)$ colors for all surfaces with $\chi < 0$ in 1968, and Appel and Haken solved the case $\chi = 2$ (the original conjecture) in 1976. However, Franklin proved in 1934 that only *six* colors are needed for maps on the Klein bottle, although the Heawood number of the Klein bottle is seven. Thus, the Heawood conjecture is true for all surfaces *except* the Klein bottle.

Chapter 6

Homology

"Mathematicians are like Frenchmen; whatever you say to them they translate into their own language and forthwith it is something completely different."

Goethe

The object of topology is the classification and description of the shape of a space up to topological equivalence. We have in Theorem 4.14 a technique for classifying the surfaces, but this is, as you may have noticed, rather arduous. The euler characteristic can be used to shorten the process, but for some cases a lengthy procedure is still necessary. Neither of these options provide a clear accounting for the ways in which the surfaces vary, e.g., which enclose cavities, which are non-orientable, etc., and neither can be completely generalized to higher-dimensional manifolds. Ideally, one would like some sort of algebraic invariant or computable quantity that would codify a lot of information: how many connected pieces a space has, how the gluing directions work, whether the surface is orientable or not, etc. The euler characteristic is a first attempt at this and has the advantage of being quite easy to compute, but it fails to distinguish between the torus and the Klein bottle, which both have $\chi = 0$. See Figure 6.1.

Fig. 6.1. The euler characteristic sees both the torus and the Klein bottle like this

The euler characteristic remembers that the edges are to be glued together, but ignores the direction of the gluing. Our next attempt must respect these directions. The gluing directions for the 1-cells of a 2-complex determine whether the 2-cell forms a hollow cavity, as in the inside of the torus, or not, as in the Klein bottle. The euler characteristic is too simple and combines too much information. For example, the set X consisting of two points (recall Fig. 5.22) has the same euler characteristic as the sphere $\mathbb{S}^2$ in spite of the obvious differences: on the 0-level, X has two components and $\mathbb{S}^2$ only one, and on the 2-cell level, X has no 2-cells, whereas $\mathbb{S}^2$ has a 2-cell which forms a hollow cavity. Our next attempt at the algebraization of topology will be a sort of layered invariant, called the homology groups, which will try to express all this information. This development requires some knowledge of abelian groups. I have tried to provide everything you need to know in the Appendix. Please take the time now to go read it, if only to get used to the notation used.

6.1 The algebra of chains

In studying the euler characteristic we counted only the number of cells of each type. In order to get more information about a complex, we want to study not only how many cells it has, but also what type of cells they are and how they are identified. Thus, we look at the cells themselves, not just how many there are.

We wish to keep track of all the cells in a complex and how they are glued together. We must also take into account the directions of any edges glued together. Mathematicians never do anything by halves, so we may as well assign directions to everything in sight.

An edge, by definition homeomorphic to the interval $[0, 1]$, has a natural direction: from initial point (corresponding to 0) to terminal point (corresponding to 1). As in Chapter 4, we denote direction along edge a by an arrow. For example, consider the complex on the torus in Figure 6.2.

Fig. 6.2. A complex on the torus

The arrow gives a natural meaning to the expression $+b$: let b stand for instructions to travel along the path labeled b in the direction of the arrow. Then $2b$ would denote a trip twice around b in the indicated direction and

$-b$ a trip along b in the backwards direction. We, thus, can assign meaning to nb for any integer $n \in \mathbb{Z}$. It is harder to make sense of $1\frac{1}{2}b$ or πb, so we will limit ourselves to the integers. Similar meanings can be assigned to multiples of the edge a. Furthermore, $a + b$ can mean a trip forwards along the a path and then forwards along b. We can define an arithmetic for edges! The quantity $6a - 2b$ means travel around the a-loop six times in the direction of the arrow, and then around b twice going against the arrow. Continuing thus, we invent geometrical interpretations for $ma + nb$, where $m, n \in \mathbb{Z}$. In the notation of the Appendix, the set of all integer combinations $ma + nb$ is

$$\{ma + nb : m, n \in \mathbb{Z}\} = [[a, b]] \simeq \mathbb{Z} \oplus \mathbb{Z}$$

Thus, a group is naturally related to the set of directed edges.

As a matter of convention, we usually use capital letters, especially P and Q, to designate vertices. Edges are assigned letters like a or b, in conformity with the labelings used in Chapter 4. For 2-cells or cells in the abstract we use Greek letters, especially σ and τ. A direction or orientation can also be assigned to 2-cells — clockwise or counterclockwise; see Figure 6.3.

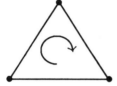 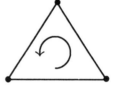

Fig. 6.3. Two orientations for 2-cells

(6.1) Definition. *A 2-complex K is directed if each edge or 1-cell is given a direction (from initial point to terminal point) and each polygon or 2-cell is given a direction (clockwise or counterclockwise).*

Note that the choices of directions for edges and polygons are arbitrary, as long as the gluing instructions for the outside edges are respected. A complex may be directed in many different ways.

We wish to study all reasonable combinations of cells and their arithmetic. Instead of trying to combine 2-cells with 1-cells and then have to keep track of which dimensions go with which cells, we will segregate them and only allow combinations of cells of like dimension.

(6.2) Definition. *Let K be a directed complex. An (integral) k-chain C in K is a sum*

$$C = a_1\sigma_1 + a_2\sigma_2 + \cdots + a_n\sigma_n$$

where $\sigma_1, \sigma_2, \ldots, \sigma_n$ are k-cells in K and $a_1, a_2, \ldots, a_n$ are integers. Define $0\sigma = \emptyset$.

Another directed complex on the torus is given in Figure 6.4.

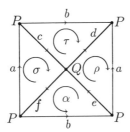

Fig. 6.4. Another complex on the torus

For the complex on the torus in Figure 6.4, some 1-chains are $2a, -b, b+c$, and $2a - b + 3c - 4f$. One can think of the 1-chain $2a - b + 3c$ as meaning instructions to travel in the forward direction along a twice, then go backwards along b, then forward along c three times, etc. Some 2-chains are $-2\sigma, \tau - 3\rho, \sigma + \tau + \rho$, etc. Think of the expression σ as meaning instructions to swirl around the cell marked σ once in the indicated direction. Then $\tau - 3\rho$ means swirl around τ in the indicated direction, then swirl around ρ three times in the direction opposite the indicator arrow.

Typical 0-chains are $2P$, $0 \cdot P = \emptyset$, $-P$, $P + Q$, $3P - 2Q$, etc. The quantity $2P$ can perhaps be thought of as instructions to hop on P twice and $P + Q$ as hop on P then on Q, but it is extremely difficult at this point to conceive of a meaningful geometric interpretation for $-P$ or $3P - 2Q$. The algebraization of topology is only achieved at the cost of some abstract nonsense.

There is a natural arithmetic for k-chains.

(6.3) Definition. *Let C and D be k-chains in a directed complex K, written as*

$$C = a_1\sigma_1 + a_2\sigma_2 + \cdots + a_n\sigma_n$$

$$D = b_1\sigma_1 + b_2\sigma_2 + \cdots + b_n\sigma_n$$

The sum $C + D$ is defined by

$$C + D = (a_1 + b_1)\sigma_1 + (a_2 + b_2)\sigma_2 + \cdots + (a_n + b_n)\sigma_n$$

For example,

$$(B - 3C) + (A + B + C) = A + 2B - 2C$$

just as one would expect, if one did not even worry about what it all meant. One only requires that A, B, and C be k-chains of the same dimension. This definition of addition obeys the usual laws.

(6.4) Theorem. *Let C_1, C_2, C_3 be any k-chains in a directed complex K.*

(1) *Commutativity:* $C_1 + C_2 = C_2 + C_1$.
(2) *Associativity:* $C_1 + (C_2 + C_3) = (C_1 + C_2) + C_3$.
(3) *Identity:* $C_1 + \emptyset = C_1$.
(4) *Inverse:* $C_1 + (-C_1) = C_1 - C_1 = \emptyset$.

In order to have property (3) be true, the null set, $\emptyset$, must be defined to be a k-chain for each value of k. Adding the null set acts like adding zero. From (4), note that a k-chain C has inverse $-C$.

▷ **Exercise 6.1.** Prove Theorem 6.4.

(6.5) Definition. *Let K be a directed complex. Denote the group of all k-chains on K by $C_k(K)$ for $k = 0, 1, \ldots, dim(K)$.*

That $C_k(K)$ is an abelian group follows from the properties listed in Theorem 6.4. Note that for finite complexes, in particular for all compact surfaces by Theorem 4.12, the groups $C_k(K)$ will be finitely generated and, thus, $C_k(K)$ will be a group of the form $\mathbb{Z}^n = \mathbb{Z} \oplus \mathbb{Z} \oplus \cdots \oplus \mathbb{Z}$.

As an example, let K be the complex on the sphere in Figure 6.5.

Fig. 6.5. Complex K on $\mathbb{S}^2$

The only 2-cell is σ. The null set $\emptyset$ is a k-chain for each k by definition. Thus,

$$C_2(K) = \{\emptyset, \sigma, 2\sigma, 3\sigma, \ldots - \sigma, -2\sigma \ldots\} = \{n\sigma : n \in \mathbb{Z}\}$$

and so $C_2(K)$ is all integer combinations of the sole 2-cell σ. In accordance with the notation of the Appendix, we recognize this group as being $C_2(K) = [[\sigma]] \simeq \mathbb{Z}$.

The 1-chains of K are

$$C_1(K) = \{\emptyset, a, \ 2a, \ldots, -a, -2a, \ldots\} = \{na : n \in \mathbb{Z}\}$$

and so $C_1(K) = [[a]] \simeq \mathbb{Z}$.

The 0-chains of K are

$$C_0(K) = \{\emptyset, P, 2P, \ldots - P, -2P, \ldots, -2Q, -Q, Q, \ 2Q, \ldots,$$
$$P + Q, P - Q, 2P + 3Q, \ldots\}$$
$$= \{nP + mQ : n, m \in \mathbb{Z}\}$$

Thus, $C_0(K)$ consists of all integer combinations of P and Q, and so $C_0(K) = [[P, Q]] \simeq \mathbb{Z} \oplus \mathbb{Z}$.

We have defined rules for determining groups, one for each dimension, which relate to the sets of all k-cells, taking into account their directions. The cells alone do not determine a complex, though. We must also keep track of which cells get glued to which and the directions of any gluings. Two 2-cells will be adjacent if they have an edge in common. Thus, which edges form the faces of a 2-cell seems to be valuable information. Similarly, edges are adjacent if they have a vertex in common, i.e. if the vertex is a face of both edges. We need a mechanism that will pick out the $(n-1)$-dimensional faces of an n-cell.

(6.6) Definition. *The boundary of a k-cell σ, denoted $\partial(\sigma)$, is the k-chain consisting of all the $(k-1)$-cells that are faces of σ, with direction inherited from the orientation of σ.*

The boundary of a 0-cell P is defined to be the null set: $\partial(P) = \emptyset$. The boundary of a directed 1-cell a as in Figure 6.6 is defined by

$$\partial(a) = Q - P = (\text{terminal point} - \text{initial point}) = (\text{head} - \text{tail}).$$

In other words, $\partial(a) = Q - P$ tells us that edge a goes connects vertex P to vertex Q, and, what is more, is given the direction that starts at P and goes toward Q.

The boundary of an oriented 2-cell σ is the chain formed by the 1-cells on the boundary of σ, with sign $+$ if the direction of an edge is consistent with the direction of the 2-cell, and $-$ otherwise. In Figure 6.7, $\partial(\sigma) = a + b - c$, since if one travels around the boundary of σ in the indicated direction (clockwise), then one will go forward on a, forward on b, and then backward along c.

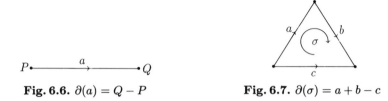

$P \bullet \xrightarrow{\quad a \quad} \bullet Q$

Fig. 6.6. $\partial(a) = Q - P$ **Fig. 6.7.** $\partial(\sigma) = a + b - c$

The boundary of an individual k-cell is thus defined. To generalize to k-chains, we must consider combinations of k-cells. In Figure 6.8, note that the 2-cells σ and τ have the same clockwise orientation, and also that they have edge b in common. Considering them separately, we have

$$\partial(\sigma) = a + b - c$$

$$\partial(\tau) = d - e - b$$

What should $\partial(\sigma + \tau)$ be? Considering the equations above, the arithmetic practically begs us to define

$$\partial(\sigma + \tau) = (a + b - c) + (d - e - b) = a - c + d - e.$$

Thus, the b-edges should cancel each other out. This makes some sense, since we can combine σ and τ to make one bigger 2-cell $\sigma + \tau$ with the clockwise orientation, which would have boundary $a - c + d - e$. If σ and τ have different orientations, it would not make sense to try to combine them into a single cell, but then the arithmetic would be different, too.

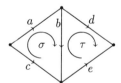

Fig. 6.8. $\partial(\sigma + \tau) = (a + b - c) + (d - e - b) = a - c + d - e$

(6.7) Definition. *Let C be a k-chain, so $C = a_1\sigma_1 + a_2\sigma_2 + \cdots + a_m\sigma_m$. The boundary of C is*

$$\partial(C) = a_1\partial(\sigma_1) + a_2\partial(\sigma_2) + \cdots + a_m\partial(\sigma_m)$$

We are allowing the demands of both the natural arithmetic and the geometry of the cells dictate the correct definition. The boundary operation takes a k-chain and gives back a $(k-1)$-chain. Thus, this operator can be considered as a function $\partial : C_k(K) \longrightarrow C_{k-1}(K)$.

▷ **Exercise 6.2.**   Consider the complex on the Klein bottle shown in Figure 6.9.

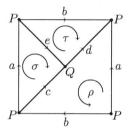

Fig. 6.9. A complex on $\mathbb{K}^2$

Trace the following chains and find their boundaries:

(1) $2a + b$
(2) $d - a$
(3) $2b - c - d$
(4) $\sigma + \tau$
(5) $\tau - \rho$
(6) $P + 2Q$

▷ **Exercise 6.3.**   Show that $\partial : C_k(K) \longrightarrow C_{k-1}(K)$ is a homomorphism for any complex K.

The purpose of all of this is to come up with something that will distinguish the shape of a complex. Important characteristics of shape include the ability to enclose a cavity, as in the sphere or torus.

In the complex on the sphere in Figure 6.10, note that the 2-cell σ has boundary $\partial(\sigma) = a - a = \emptyset$. Similarly, on the torus of Figure 6.11, $\partial(\tau) = a + b - a - b = \emptyset$. Geometrically, $\partial(\sigma) = \emptyset$ does not mean that the cell has no boundary, but rather that the boundary edges cancel.

The 1-cells that form loops, starting and ending at the same vertex, seem to be important. Recall Chapter 5.2, where the presence of such loops determined the euler characteristic of a graph. On the torus of Figure 6.11, both a and b form loops, and $\partial(a) = \partial(b) = P - P = \emptyset$.

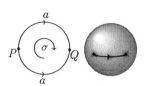

Fig. 6.10. Sphere

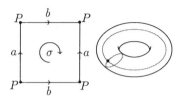

Fig. 6.11. Torus

(6.8) Definition. *If C is a k-chain in a directed complex K and $\partial(C) = \emptyset$, then C is a k-cycle. The set of all k-cycles in K is $Z_k(K) \subseteq C_k(K)$.*

▷ **Exercise 6.4.** Prove that $Z_k(K)$ is a subgroup of $C_k(K)$ and that $Z_k(K)$ is the kernel of $\partial : C_k(K) \longrightarrow C_{k-1}(K)$.

On the other hand, in the Klein bottle in Figure 6.12 the 2-cell σ fails to enclose a cavity, and $\partial(\sigma) = a + b - a + b = 2b$. It might be nice to keep track of the quantity $2b$, the boundary of σ, as a measure of how σ failed to form a cavity. Note that in the sphere in Figure 6.10, the boundary of the 1-cell a is $Q - P$, and so a fails to form a loop.

Fig. 6.12. Klein bottle

(6.9) Definition. *If C is a k-cycle in a directed complex K such that there exists a $(k+1)$-chain D with $\partial(D) = C$, then C is a k-boundary. The set of all k-boundaries in K is $B_k(K) \subseteq C_k(K)$.*

▷ **Exercise 6.5.** Prove that $B_k(K)$ is a subgroup of $C_k(K)$ and that $B_k(K)$ is the image $R(\partial)$ of $\partial : C_{k+1}(K) \longrightarrow C_k(K)$.

Consider the torus complex of Figure 6.13. Neither σ nor τ are 2-cycles, since $\partial(\sigma) = a - c - b$ and $\partial(\tau) = a - b - c$; however,

$$\partial(\sigma - \tau) = (a - b - c) - (a - b - c) = \emptyset$$

so $\sigma - \tau$ is a 2-cycle, as is any multiple of $\sigma - \tau$. Thus,

$$Z_2(K) = \{\ldots, -2(\sigma - \tau), -(\sigma - \tau), \emptyset, (\sigma - \tau), 2(\sigma - \tau), \ldots\}$$
$$= \{n(\sigma - \tau) : n \in \mathbb{Z}\} = [[(\sigma - \tau)]] \simeq \mathbb{Z}$$

There are, of course, no 2-boundaries since there are no 3-cells for 2-chains to be boundaries of, so

$$B_2(K) = \{\emptyset\} \simeq 0$$

Fig. 6.13. A complex on $\mathbb{T}^2$

Of the 1-chains, a is a 1-cycle since $\partial(a) = P - P = \emptyset$, and so are b and c. Then $a + b$ is also a 1-cycle, and so are $a - 3b$, $47a + 64c$, $2a - 3b + c$, etc. Any combination of a, b, and c will be a 1-cycle:

$$Z_1(K) = [[a, b, c]] \simeq \mathbb{Z} \oplus \mathbb{Z} \oplus \mathbb{Z}$$

Since $\partial(\sigma) = \partial(\tau) = a - b - c$, the 1-chain $a - b - c$ is a 1-boundary. Similarly, $\partial(3\sigma) = 3(a - b - c)$, etc., so $B_1(K)$ is the set of all multiples of $(a - b - c)$:

$$B_1(K) = \{n(a - b - c) : n \in \mathbb{Z}\} = [[(a - b - c)]] \simeq \mathbb{Z}$$

Considering the 0-chains, note that the vertex P is a 0-cycle, since $\partial(P) = \emptyset$:

$$Z_0(K) = [[P]] \simeq \mathbb{Z}$$

Since $\partial(a) = \partial(b) = \partial(c) = P - P = \emptyset$, the 1-cells of the complex have null boundaries. Thus, no multiple of P is a 0-boundary, so

$$B_0(K) = \{\emptyset\} \simeq 0$$

▷ **Exercise 6.6.** For the complex of Exercise 6.1, find $Z_k(K)$ and $B_k(K)$, for $k = 0, 1, 2$.

As in Chapter 4, there often advantages (usually theoretical and not practical) in using simplicial complexes, which have only triangular cells.

For example, a k-simplex must always have $k+1$ vertices, but in a non-simplicial complex, a polygon can have any number (greater than two) of vertices. Thus, the number of vertices tells one immediately the dimension of the simplex. Further, the ordering of the vertices gives a natural orientation on the simplex. Let us denote by $\langle v_0 \rangle$ the vertex or 0-simplex v_0; $\langle v_0, v_1 \rangle$ is the edge or 1-simplex running from v_0 to v_1; $\langle v_0, v_1, v_2 \rangle$ is the 2-simplex with vertices v_0, v_1, v_2 (we are assuming that these vertices are not collinear) oriented in the order of these vertices, as in Figure 6.14. A 3-simplex $\langle v_0, v_1, v_2, v_3 \rangle$ is directed by a helix or screw oriented by the order of the vertices.

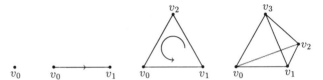

Fig. 6.14. The ordering of the vertices induces an orientation on the simplex

An n-simplex is denoted by $\langle v_0, v_1, v_2, \dots, v_n \rangle$. Changing the order of the vertices can change the orientation of the simplex. In Figure 6.14, note $\langle v_1, v_2, v_0 \rangle = \langle v_0, v_1, v_2 \rangle$ but $\langle v_1, v_0, v_2 \rangle = -\langle v_0, v_1, v_2 \rangle$. In general, if one interchanges two vertices, the orientation is reversed.

$$-\langle v_0, v_1, v_2, \dots, v_n \rangle = \langle v_1, v_0, v_2, \dots, v_n \rangle$$

Note that it takes two interchanges to change the example above:

$$\langle v_1, v_2, v_0 \rangle = -\langle v_1, v_0, v_2 \rangle = +\langle v_0, v_1, v_2 \rangle$$

The formula gives the same orientations as the intuitive geometry of the simplex.

The 2-simplex $\langle v_0, v_1, v_2 \rangle$ of Figure 6.14 has three edges: $\langle v_0, v_1 \rangle$, $\langle v_1, v_2 \rangle$, and $\langle v_2, v_0 \rangle = -\langle v_0, v_2 \rangle$. These are formed by choosing any two of the three vertices.

$$\partial(\langle v_0, v_1, v_2 \rangle) = \langle v_1, v_2 \rangle - \langle v_0, v_2 \rangle + \langle v_0, v_1 \rangle$$

The faces of an n-simplex $\langle v_0, v_1, v_2, \dots, v_n \rangle$ will be all the $(n-1)$-simplices formed by these vertices. These $(n-1)$-simplices will have n vertices chosen from $v_0, v_1, v_2, \dots, v_n$, so we can designate one by $\langle v_0, v_1, v_2, \dots, \overline{v_k}, \dots, v_n \rangle$ where $\overline{v_k}$ means omit the vertex v_k. The boundary of an n-simplex is

$$\partial(\langle v_0, v_1, v_2, \dots, v_n \rangle) = \sum_{k=0}^{n} (-1)^k \langle v_0, v_1, v_2, \dots, \overline{v_k}, \dots, v_n \rangle$$

These rules can simplify the proofs of some theorems, especially in dimensions greater than 2, when pictures are difficult to draw and can be misleading.

(6.10) Theorem. *The composition $\partial \circ \partial : C_k(K) \longrightarrow C_{k-2}(K)$ satisfies $\partial \circ \partial(C) = \emptyset$ for any k-chain C on the complex K. That is, $\partial \circ \partial = 0$.*

(6.11) Lemma. *If σ is a n-simplex, then $\partial \circ \partial(\sigma) = \emptyset$.*

Proof. As noted above, σ may be represented by

$$\sigma = \langle v_0, v_1, v_2, \ldots, v_k, \ldots, v_n \rangle.$$

Then

$$\partial \circ \partial(\sigma) = \partial \left(\sum_{k=0}^{n} (-1)^k \langle v_0, v_1, v_2, \ldots, \overline{v_k}, \ldots, v_n \rangle \right)$$

$$= \sum_{k=0}^{n} (-1)^k \partial \big(\langle v_0, v_1, v_2, \ldots, \overline{v_k}, \ldots, v_n \rangle \big)$$

$$= \sum_{k=0}^{n} (-1)^k \left[\sum_{\ell=0}^{k-1} (-1)^\ell \langle v_0, v_1, \ldots, \overline{v_\ell}, \ldots, \overline{v_k}, \ldots, v_n \rangle \right.$$

$$\left. + \sum_{\ell=k+1}^{n} (-1)^{\ell-1} \langle v_0, v_1, \ldots, \overline{v_k}, \ldots, \overline{v_\ell}, \ldots, v_n \rangle \right]$$

$$= \sum_{\ell < k \leq n} (-1)^{k+\ell} \langle v_0, v_1, \ldots, \overline{v_\ell}, \ldots, \overline{v_k}, \ldots, v_n \rangle$$

$$+ \sum_{k < \ell \leq n} (-1)^{k+\ell-1} \langle v_0, v_1, \ldots, \overline{v_k}, \ldots, \overline{v_\ell}, \ldots, v_n \rangle$$

$$= \sum_{k \neq \ell} \left[(-1)^{k+\ell} + (-1)^{k+\ell-1} \right] \langle v_0, v_1, \ldots, \overline{v_k}, \ldots, \overline{v_\ell}, \ldots, v_n \rangle$$

$$= \sum_{k \neq \ell} 0 \cdot \langle v_0, v_1, \ldots, \overline{v_k}, \ldots, \overline{v_\ell}, \ldots, v_n \rangle$$

$$= \emptyset$$

$\square$

▷ **Exercise 6.7.** Let σ be a standard k-simplex for $k = 0, 1, 2$. Verify Lemma 6.11 geometrically for these cases.

Proof of Theorem 6.10. We assume that the complex is triangulated. Let $C = a_1 \sigma_1 + a_2 \sigma_2 + \cdots + a_n \sigma_n$ where a_i is an integer and σ_i is a k-simplex in K.

$$\partial(C) = \partial(a_1 \sigma_1 + a_2 \sigma_2 + \cdots + a_n \sigma_n)$$

$$
\begin{aligned}
&= \partial(a_1\sigma_1) + \partial(a_2\sigma_2) + \cdots + \partial(a_n\sigma_n) \\
&= a_1\partial(\sigma_1) + a_2\partial(\sigma_2) + \cdots + a_n\partial(\sigma_n) \\
\partial \circ \partial(C) &= \partial(\partial(C)) \\
&= \partial[a_1\partial(\sigma_1) + a_2\partial(\sigma_2) + \cdots + a_n\partial(\sigma_n)] \\
&= a_1\partial(\partial(\sigma_1)) + a_2\partial(\partial(\sigma_2)) + \cdots + a_n\partial(\partial(\sigma_n)) \\
&= a_1\partial \circ \partial(\sigma_1) + a_2\partial \circ \partial(\sigma_2) + \cdots + a_n\partial \circ \partial(\sigma_n)
\end{aligned}
$$

The proof then follows from Lemma 6.11. $\hfill\square$

▷ **Exercise 6.8.** Show that $B_k(K)$ is a subgroup of $Z_k(K)$.

6.2 Homology

The trouble with the groups C_k, Z_k, and B_k is that they depend too much on the complex and so obscure the underlying space. The chain groups, C_k, list all the cells of K, whereas the boundary homomorphism, in describing which cells form the boundary of a given cell and the direction of those cells, tells how the cells are glued together to form the complex, but this is, in a way, too much information. We need some way of eliding the extraneous data to pick out only the essential facts about the shape of the figure. Two different complexes on the torus, for example Figures 6.2 and 6.4, would give totally different chain groups. We wish somehow to say that these have the same shape or underlying surface.

Note that a 2-cycle can be described as a chain whose edges are gathered together to form a hollow cavity. The 1-cycles are edges which form loops. A chain which is not a cycle cannot enclose a cavity or loop. Thus, the cycles seem to be of especial importance in determining the shape of the object, more so than the general chain groups. On the other hand, some cycles seem to be redundant.

In Figure 6.15, the chain $c + d$ starts and ends at the same points that b does, and does not seem to convey any information about the shape that the cycles a and b do not already tell us. We would like some legitimate way to get rid of $c + d$ algebraically.

We are trying to accomplish algebraically what Theorem 4.14 did geometrically: to find ways of screening out extraneous information. In Theorem 4.14, all the extra cells and edges from the original triangulation were eliminated to end up with one of the simple standard planar diagrams. To try to excerpt the desired information, we define an equivalence relation on chains.

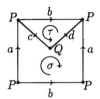

Fig. 6.15. A complex on the torus

(6.12) Definition. *The k-chains C_1 and C_2 are homologous, written $C_1 \sim C_2$, if $C_1 - C_2 \in B_k(K)$; i.e., if $C_1 - C_2 = \partial(D)$ for some $(k+1)$-chain D.*

In Figure 6.15, $c + d \sim b$, since $c + d - b = \partial(\tau)$. In a sense, $c + d$ is not needed, as long as we have b. The 2-cell τ forms a bridge between the edges $c + d$ and b, so any information carried by $c + d$ is also carried by b, and vice versa. Similarly, $\partial(c) = Q - P$, so $Q \sim P$. The edge c forms a bridge from Q to P. Note that $\partial(\sigma) = a + c + d - a - b = c + d - b$, so $c + d \sim b$, which we already knew, and $\partial(d) = P - Q$, so $P \sim Q$. We get no additional information from those computations.

(6.13) Theorem. *Let K be a complex, with C_1, C_2, C_3, C_4 chains in $C_k(K)$:*

(1) $C_1 \sim C_1$.
(2) *If $C_1 \sim C_2$, then $C_2 \sim C_1$.*
(3) *If $C_1 \sim C_2$ and $C_2 \sim C_3$, then $C_1 \sim C_3$.*
(4) *If $C_1 \sim C_2$ and $C_3 \sim C_4$, then $C_1 + C_3 \sim C_2 + C_4$.*

▷ **Exercise 6.9.** Prove Theorem 6.13.

Conditions (1)–(3) verify that homology is an equivalence relation, and (4) guarantees that it works well in combination with the chain addition of Definition 6.3. Thus, $\sim$ acts like $=$ and can be used in place of it.

Note that two 0-cells P and Q will be homologous if there is an edge a connecting P to Q, so that $\partial(a) = P - Q$ and $P \sim Q$. Thus, P and Q are homologous if and only if there is a path on the complex between them, which will be true if they are in the same component. For example, consider the complex of Figure 6.16.

$$\partial(a) = Q - P \text{ so } P \sim Q, \qquad \partial(d) = S - R \text{ so } S \sim R$$
$$\partial(b) = T - P \text{ so } P \sim T, \qquad \partial(h) = V - U \text{ so } U \sim V$$
$$\partial(c) = R - Q \text{ so } R \sim Q, \qquad \partial(i) = W - V \text{ so } W \sim V$$

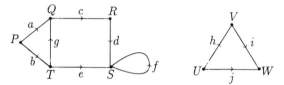

Fig. 6.16. $P \sim Q \sim R \sim S \sim T$ and $U \sim V \sim W$

Thus, $P \sim Q \sim R \sim S \sim T$ and $U \sim V \sim W$ but $P \not\sim U$. Up to homology there are only two vertices, P and U, one for each connected component.

A 1-cycle must have $\partial(C) = \emptyset$, so that $\partial(C)$ must be something like $\partial(C) = P - P = \emptyset$. Thus, C must form a loop. A 1-cycle C is *trivial* if and only if its interior is filled by a 2-chain σ with $C = \partial(\sigma)$ and, therefore, $C \sim \emptyset$. Figures 6.17 and 6.18 represent a non-trivial 1-cycle and a trivial 1-cycle, respectively.

Fig. 6.17. $a \not\sim \emptyset$ **Fig. 6.18.** $b \sim \emptyset$

Two loops are homologous if they form the boundary of a 2-chain. In Figure 6.19, $\partial(\sigma) = a + c - b - c = a - b$, so $a \sim b$. Thus, the loops are homologous if there is a 2-cell forming a bridge between them.

A 2-cycle must have its edges glued together to enclose a cavity, as in the torus or the sphere. A 2-cycle can only be trivial in homology if it is the boundary of some 3-chain. For example, in the solid ball, the skin, a sphere, is homologous to $\emptyset$. Remember that the cycles seem to be the most important chains in determining shape.

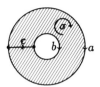

Fig. 6.19. $a \sim b$

(6.14) Definition. *Let K be a directed complex. The kth homology group of K is $H_k(K) = Z_k(K)/B_k(K)$, the group of equivalence classes of elements of $Z_k(K)$ with the homology relation. In other words, $H_k(K)$ is $Z_k(K)$ with homology used instead of equality.*

The homology groups combine all the essential information we have gathered: which cells form cavities or loops; and cancels out the extraneous information. The process of finding $H_k(K)$ is outlined below:

(6.15) Computation of $H_k(K)$ for a directed complex K.
First label and indicate orientation of all cells of the complex. Do the computations for one dimension at a time, starting with the highest.

(1) Find $C_k(K)$, the group of all k-chains.
(2) For each generating k-chain C from (1), compute $\partial(C)$.
(3) Find $Z_k(K)$, using your computations from (2).
 (a) Note that if $Z_k(K) = \{\emptyset\}$, then $H_k(K) = \{\emptyset\}$.
(4) Find $B_k(K)$.
 (a) When working in the highest dimension, note that $B_k(K) = \emptyset$, since there are no $(k+1)$-chains for which the k-chains can form boundaries.
 (b) Otherwise, look back to (2) in the next highest dimension, where you already computed which k-chains are boundaries of $(k+1)$-cells.
 (c) Note that if $B_k(K) = \{\emptyset\}$, then $H_k(K) = Z_k(K)$.
(5) Compute $H_k(K)$, by taking $Z_k(K)$ from (3) and using any homologies found in (4).

This may seem quite intimidating at first, but with practice it is not so bad; definitely preferable to Theorem 4.14, anyhow. As a general rule when forming $H_k(K)$, the generators come from $Z_k(K)$ and the relations from $B_k(K)$.

Examples of the computation of the homology groups for some familiar surfaces follow.

(6.16) Example. Let us compute the homology groups for the sphere $\mathbb{S}^2$ with the complex given by the standard planar diagram (Fig. 6.20).

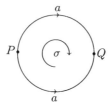

Fig. 6.20. The complex K on $\mathbb{S}^2$

Since there are 0-, 1-, and 2-cells present, we must compute $H_0(K)$, $H_1(K)$, and $H_2(K)$.

First find $C_2(K)$:

The only 2-cell is σ. The 2-chains are $\sigma, 2\sigma, 3\sigma, \ldots, -\sigma, -2\sigma, \ldots$ and $\emptyset$, so $C_2(K) = [[\sigma]] \simeq \mathbb{Z}$.

Find $Z_2(K)$:

Note that σ is a 2-cycle, since $\partial(\sigma) = a - a = \emptyset$, as are 2σ, $-\sigma$, 3σ, -47σ, etc., so $Z_2(K) = [[\sigma]] \simeq \mathbb{Z}$.

Find $B_2(K)$:

There are no 3-cells, so σ cannot be the boundary of anything, and $B_2(K) = \{\emptyset\} \simeq 0$.

Compute $H_2(K)$:

The only 2-cycles are $\emptyset$ and multiples of σ.

Note $\sigma - \emptyset = \sigma \notin B_2(K) = \{\emptyset\}$, so $\sigma \not\sim \emptyset$. Therefore,

$$H_2(K) = Z_2(K) = [[\sigma]] \simeq \mathbb{Z}$$

Repeat for $H_1(K)$:

$C_1(K) = \{a, 2a, \ldots, \phi, -a, -2a, \ldots\} = [[a]] \simeq \mathbb{Z}$.

$\partial(a) = Q - P$ so a is *not* a 1-cycle, nor are $2a$, $-a$, etc., so there are no non-trivial 1-cycles. Thus, $Z_1(K) = \{\emptyset\} \simeq 0$.

Since $\partial(\sigma) = \emptyset$, only $\emptyset$ is a 1-boundary, so $B_1(K) = \{\emptyset\} \simeq 0$.

$$H_1(K) = Z_1(K) = \{\emptyset\} \simeq 0$$

Repeat for $H_0(K)$:

$C_0(K) = \{\emptyset, P, 2P, \ldots -P, \ldots, -Q, Q, 2Q, \ldots, P+Q, 2P-3Q, \text{etc.}\}$, so $C_0(K) = [[P, Q]] \simeq \mathbb{Z} \oplus \mathbb{Z}$.

$\partial(P) = \emptyset$ and $\partial(Q) = \emptyset$ so P and Q and all combinations are 0-cycles.

$Z_0(K) = [[P, Q]] \simeq \mathbb{Z} \oplus \mathbb{Z}$.

$\partial(a) = Q - P$ so $Q \sim P$. Therefore, $2Q \sim 2P$, $-Q \sim -P$, and $P + Q \sim P + P = 2P$, etc., and so any combination of the vertices P and Q is homologous to some multiple of P.

$$H_0(K) = \{\emptyset, P, 2P, \dots, -P, -2P, \dots\} = [[P]] \simeq \mathbb{Z}$$

(6.17) Example. We can compute the homology groups of the n-dimensional sphere $\mathbb{S}^n$ for $n > 1$ by using the complex K' consisting of an n-disc σ with the boundary of σ gathered together at a point denoted P. For example, this complex for $\mathbb{S}^2$ is pictured in Figure 6.21.

Fig. 6.21. Another complex K' on $\mathbb{S}^2$

H_n: $C_n(K') = [[\sigma]] \simeq \mathbb{Z}$.
$\qquad \partial(\sigma) = \emptyset$, since $\dim(P) < n - 1$; thus σ is a cycle.
$\qquad Z_n(K') = [[\sigma]] \simeq \mathbb{Z}$.
$\qquad B_n(K') = \{\emptyset\}$ since there are no $(n+1)$-cells.

$$H_n(K') = Z_n(K') = [[\sigma]] \simeq \mathbb{Z}$$

H_i: $1 \le i < n$: $C_i(K') = \{\emptyset\}$, so

$$H_i(K') = \{\emptyset\} \simeq 0$$

H_0: $C_0(K') = [[P]] \simeq \mathbb{Z}$.
$\qquad \partial P = \emptyset$ so P is a cycle and $Z_0(K') = [[P]] \simeq \mathbb{Z}$.
$\qquad$ Since there are no 1-cells, $B_0(K') = \{\emptyset\}$.

$$H_0(K') = Z_0(K') = [[P]] \simeq \mathbb{Z}$$

These computations give the same homology groups as Example 6.16 for the 2-sphere, and in addition we have computed the homology groups for all higher-dimensional spheres.

(6.18) Example. The $(n+1)$-dimensional solid ball

$$\mathbb{B}^{n+1} = \{\mathbf{x} \in \mathbb{R}^{n+1} : \|\mathbf{x}\| \le 1\}$$

is an $(n+1)$-manifold with boundary the n-sphere $\mathbb{S}^n$. By a slight abuse of notation we write

$$\partial(\mathbb{B}^{n+1}) = \mathbb{S}^n = \{\mathbf{x} \in \mathbb{R}^{n+1} : \|\mathbf{x}\| = 1\}$$

We can compute the homology groups of the $(n+1)$-dimensional ball $\mathbb{B}^{n+1}$ for $n > 2$ by using the complex K' of Example 6.17 with the addition of an $(n+1)$-cell which fills the inside of the n-sphere. Thus, we let the complex K consist of an $(n+1)$-cell τ with boundary the n-cell σ, and the boundary of σ gathered together at a point denoted P.

H_{n+1}: $C_{n+1}(K) = [[\tau]] \simeq \mathbb{Z}$.
$\qquad \partial(\tau) = \sigma$, so τ is not a cycle. Therefore, $Z_{n+1}(K) = \{\emptyset\}$ and

$$H_{n+1}(K) = \{\emptyset\} \simeq 0$$

H_n: $C_n(K) = [[\sigma]] \simeq \mathbb{Z}$.
$\qquad \partial(\sigma) = \emptyset$, since $\dim(P) < n - 1$; thus σ is a cycle.
$\qquad Z_n(K) = [[\sigma]] \simeq \mathbb{Z}$.
$\qquad$ Since $\partial(\tau) = \sigma$, it follows that $\sigma \sim \emptyset$.

$$H_n(K) = \{\emptyset\} \simeq 0$$

H_i: $1 \leq i < n$: $C_i(K) = \{\emptyset\}$, so

$$H_i(K) = \{\emptyset\} \simeq 0$$

H_0: $C_0(K) = [[P]] \simeq \mathbb{Z}$.
$\qquad \partial P = \emptyset$ so P is a cycle and $Z_0(K) = [[P]] \simeq \mathbb{Z}$.
$\qquad$ Since there are no 1-cells, $B_0(K) = \{\emptyset\}$.

$$H_0(K) = Z_0(K) = [[P]] \simeq \mathbb{Z}$$

For the case $n = 1$, we have $\partial(\mathbb{B}^2) = \mathbb{S}^1$. In Exercise 6.10, you will compute the homology groups for this disc and for the circle. For $n = 0$, note that $\mathbb{B}^1$ is the line segment $[-1, 1]$. Then $\partial(\mathbb{B}^1) = \{-1, 1\}$. We can define the 0-dimensional sphere, $\mathbb{S}^0$, to be the set of two points, so that $\partial(\mathbb{B}^{n+1}) = \mathbb{S}^n$ in all dimensions. It is easy to show that $H_1(\mathbb{B}^1) \simeq 0$ and $H_0(\mathbb{B}^1) \simeq \mathbb{Z}$. One can define the 0-dimensional ball $\mathbb{B}^0$ to be a single point. Thus, for all dimensions $n \geq 0$ we have

$$H_k(\mathbb{B}^{n+1}) \simeq \begin{cases} 0, & 1 \leq k \leq n+1 \\ \mathbb{Z}, & k = 0 \end{cases}$$

In all dimensions except $n = 0$,

$$H_k(\mathbb{S}^n) \simeq \begin{cases} \mathbb{Z}, & k = n \\ 0, & 0 < k < n \\ \mathbb{Z}, & k = 0 \end{cases}$$

For the case $n = 0$, note that $H_0(\mathbb{S}^0) \simeq \mathbb{Z} \oplus \mathbb{Z}$.

(6.19) Example. $\mathbb{K}^2$, the Klein bottle, with complex K'' of Figure 6.22.

Fig. 6.22. The complex K'' on $\mathbb{K}^2$

H_2: $C_2(K'') = [[\sigma]] \simeq \mathbb{Z}$.
    $\partial(\sigma) = c + d - c + d = 2d$ so σ is a not a cycle, and $Z_2(K'') = \{\emptyset\} \simeq 0$.
    $B_2(K'') = \{\emptyset\}$ since there are no 3-cells.

$$H_2(K'') = Z_2(K'') \simeq 0$$

H_1: $C_1(K'') = [[c, d]] \simeq \mathbb{Z} \oplus \mathbb{Z}$.
    $\partial(c) = \partial(d) = Q - Q = \emptyset$, so $Z_1(K'') = [[c, d]] \simeq \mathbb{Z} \oplus \mathbb{Z}$.
    Since $\partial(\sigma) = 2d$, in the homology group $2d \sim \emptyset$.

$$H_1(K'') = [[c, d : 2d = 0]] \simeq \mathbb{Z} \oplus \mathbb{Z}/2$$

H_0: $C_0(K'') = [[Q]] \simeq \mathbb{Z}$.
    $\partial Q = \emptyset$ so Q is a cycle and $Z_0(K'') = [[Q]] \simeq \mathbb{Z}$.
    Since $\partial(c) = \emptyset$ and $\partial(d) = \emptyset$, $B_0(K'') = \{\emptyset\}$.

$$H_0(K'') = Z_0(K'') = [[Q]] \simeq \mathbb{Z}$$

With practice, the groups are not difficult to compute, certainly less tedious than the process of Theorem 4.14. The next exercise should provide plenty of practice.

▷ **Exercise 6.10.**   Compute the integral homology groups for the following spaces:

  (1) $\mathbb{P}^2$
  (2) $\mathbb{T}^2$
  (3) the disc
  (4) the cylinder
  (5) the Möbius band
  (6) the circle, $\mathbb{S}^1$
  (7) a tree of your choice
  (8) the double banana of Figure 6.23
  (9) a sphere with two whiskers, as illustrated in Figure 6.24

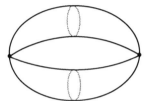   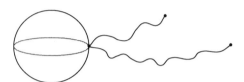

Fig. 6.23. Double banana   **Fig. 6.24.** Sphere with two whiskers

  From the computations of Example 6.19 and Exercise 6.10(2), note that $H_2(\mathbb{T}^2) = \mathbb{Z}$ and $H_2(\mathbb{K}^2) = 0$. The homology groups distinguish between the torus and the Klein bottle, whereas the euler characteristic did not.

  In each dimension, $H_k(K)$ gives some piece of information relating to properties determined by that dimension. The group $H_0(K)$ measures the connectivity of the complex K, since for vertices P and Q, $P \sim Q$ if and only if there is an 1-chain connecting them. In general, $H_0(K) = \mathbb{Z}^n$, where n is the number of connected components of K. The group $H_1(K)$ counts non-trivial loops and $H_2(K)$ counts hollows or cavities in a 2-dimensional figure. For the sphere of Example 6.16, $H_2(\mathbb{S}^2) = \mathbb{Z}$, but from Example 6.19, note that $H_2(\mathbb{K}^2) = 0$. This is because the Klein bottle does not enclose a cavity. Only the orientable surfaces can enclose cavities.

(6.20) Theorem. *Let S be a compact connected surface without boundary. If S is orientable, then $H_2(S) \simeq \mathbb{Z}$. If S is not orientable, then $H_2(S) \simeq 0$.*

Proof. We assume that the surface S has a triangulation K. Let σ and τ be two adjacent 2-simplices with common edge a. These have the same orientation if they are oriented so that the edge a is canceled when forming the sum $\sigma + \tau$, as in Figure 6.25. When the cells have opposite orientations, note that $\pm 2a \in \partial(\sigma + \tau)$, as in Figure 6.26.

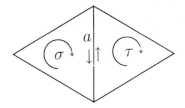

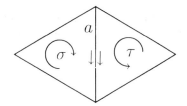

Fig. 6.25. These cells have the same orientation

Fig. 6.26. These cells have opposite orientations

Note that the 2-simplices of K cannot be oriented the same if and only if $|K|$ contains a Möbius band; see Figure 6.27.

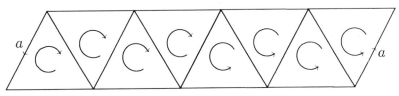

Fig. 6.27. The Möbius band cannot be oriented

If all the 2-simplices of K can be directed so that they have the same orientation, then K must be a complex on an orientable surface, either the sphere or a connected sum of tori. Consider the 2-chain C consisting of all the 2-simplices in K. Since each edge in K occurs as the boundary of precisely two triangles by Definition 4.11, and the triangles have been oriented compatibly, $\partial(C) = \emptyset$ and C is a cycle. Since there are no 3-cells, the second homology group must be $H_2(K) = [[C]] \simeq \mathbb{Z}$.

If K contains a Möbius band, then K is a complex on one of the non-orientable surfaces, which are connected sums of projective planes. Let C be a 2-cycle on K. If σ is a triangle in C, then all the triangles adjacent to σ must also be in C since $\partial(C) = \emptyset$, and so the edges of σ must be canceled by some other cell. Thus, C must contain all the 2-cells of K. However, since these cells cannot be oriented compatibly, some edges fail to cancel, as in Figure 6.26. Therefore, C is not a cycle, so $H_2(K) \simeq 0$.   □

Another peculiar thing occurs in the homology of the Klein bottle. Recall from Chapter 4.2 that $\mathbb{K}^2 = \mathbb{P}^2 \# \mathbb{P}^2$, and so is represented by two different planar diagrams, as in Example 6.19 above or as $\mathbb{P}^2 \# \mathbb{P}^2$ in Figure 6.28. Both of these complexes give $H_2(\mathbb{K}^2) \simeq \{0\}$, since $\mathbb{K}^2$ is non-orientable, and both give $H_0(\mathbb{K}^2) \simeq \mathbb{Z}$, since $\mathbb{K}^2$ is connected. By Example

6.19, $H_1(\mathbb{K}^2) = [[c, d : 2d = 0]] \simeq \mathbb{Z} \oplus \mathbb{Z}/2$ for the complex on the left in Figure 6.28. If $H_1(\mathbb{K}^2)$ is computed using the complex on the right, we get $H_1(\mathbb{K}^2) = [[e, f : 2(e + f) = 0]]$. Thus, there are apparently two different groups for $H_1(\mathbb{K}^2)$, depending on which complex is used.

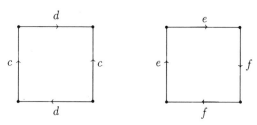

Fig. 6.28. $\mathbb{K}^2 = \mathbb{P}^2 \# \mathbb{P}^2$

That these are, after all, isomorphic is shown by the homomorphism

$$F : [[c, d : 2d = 0]] \longrightarrow [[e, f : 2(e + f) = 0]]$$

defined for the generators by

$$F(c) = e$$
$$F(d) = e + f$$

Any element $x \in [[c, d : 2d = 0]]$ can be written as $x = nc + md$ for some $n, m \in \mathbb{Z}$. Define $F(x)$ by

$$\begin{aligned} F(x) &= F(nc + md) \\ &= nF(c) + mF(d) \\ &= n(e) + m(e + f) \\ &= (n + m)e + mf \end{aligned}$$

Thus, $F(x)$ can be written as a combination of e's and f's, the generators of the second group. Note that $0 = F(2d) = 2F(d) = 2(e + f)$, so F is well-defined. F has inverse

$$G : [[e, f : 2(e + f) = 0]] \longrightarrow [[c, d : 2d = 0]]$$

defined for the generators by

$$G(e) = c$$
$$G(f) = d - c$$

Note $0 = G(2(e + f)) = 2(G(e + f)) = 2G(e) + 2G(f) = 2c + 2(d - c) = 2d$, so G is well-defined. Since F is invertible, F is an isomorphism and so these are two different presentations of the group $\mathbb{Z} \oplus \mathbb{Z}/2$. The problem of deciding when two presentations (lists of generators and relations) represent the same group is a difficult, often unsolvable, one.

The invariance of the homology groups (i.e., that they depend only on the surface and not on the complex chosen to represent that surface) will not be proven until Theorem 8.21. Even then, we only claim that topologically equivalent surfaces will have isomorphic homology groups, not necessarily in the same form.

6.3 More computations

At this point, we have figured out the homology groups of many complexes. Table 6.1 is a listing of homology groups computed thus far, either as examples or as exercises.

Table 6.1. Homology groups

Manifold	H_2	H_1	H_0
$\mathbb{S}^2$	$\mathbb{Z}$	0	$\mathbb{Z}$
$\mathbb{T}^2$	$\mathbb{Z}$	$\mathbb{Z} \oplus \mathbb{Z}$	$\mathbb{Z}$
$\mathbb{K}^2$	0	$\mathbb{Z} \oplus \mathbb{Z}/2$	$\mathbb{Z}$
$\mathbb{P}^2$	0	$\mathbb{Z}/2$	$\mathbb{Z}$
$\mathbb{S}^1$	0	$\mathbb{Z}$	$\mathbb{Z}$
cylinder	0	$\mathbb{Z}$	$\mathbb{Z}$
Möbius band	0	$\mathbb{Z}$	$\mathbb{Z}$
disc	0	0	$\mathbb{Z}$

Note that different complexes, for example the circle and the cylinder, can have the same homology groups. New complexes can be constructed with any desired homology groups, as long as they are combinations of the groups above, by using the theorems below:

(6.21) Theorem. *Let K and L be connected cell complexes with $K \cap L = \emptyset$. Let $X = K \cup L$. Then for $k = 0, 1, 2, \ldots$*

$$H_k(X) = H_k(K) \oplus H_k(L)$$

The space $X = K \cup L$ so formed is called the disjoint union of K and L.

▷ **Exercise 6.11.**   Prove Theorem 6.21.

(6.22) Theorem. *Let K and L be connected cell complexes with $K \cap L = \{P\}$, for a vertex P. Let $X = K \cup L$. Then X is a cell complex and*

$$H_k(X) = H_k(K) \oplus H_k(L), \qquad k > 0$$
$$H_0(X) = \mathbb{Z}, \qquad\qquad\quad k = 0$$

The space X so formed is called the wedge product or one-point union and is denoted $X = K \vee L$. The space $K \vee L$ is formed by joining K to L at a single point.

▷ **Exercise 6.12.**   Prove Theorem 6.22.

If one wishes to construct a complex K with integral homology groups $H_2(K) = \mathbb{Z} \oplus \mathbb{Z}$, $H_1(K) = \mathbb{Z}$, $H_0(K) = \mathbb{Z} \oplus \mathbb{Z}$, then K must have two connected components, enclose two cavities, and have one non-trivial loop. Thus, K can be any of those shown in Figure 6.29 or a similar complex.

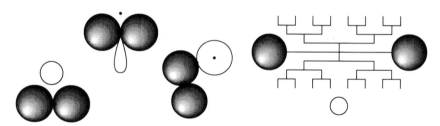

Fig. 6.29. All of these have $H_2(K) = \mathbb{Z} \oplus \mathbb{Z}$, $H_1(K) = \mathbb{Z}$, $H_0(K) = \mathbb{Z} \oplus \mathbb{Z}$

If one were to construct a complex with $H_2(K) = 0$, $H_1(K) = \mathbb{Z} \oplus \mathbb{Z}/2$, and $H_0(K) = \mathbb{Z}$, one can use either $K = \mathbb{K}^2$ or $K = \mathbb{P}^2 \vee \mathbb{S}^1$. Only the first is a surface. The homology groups of a complex cannot tell if the space is a surface or not. On the other hand, if one knows that a complex is a surface and one knows the homology groups of the space, one can identify which surface it is (up to topological equivalence) by using the results of this chapter.

One can also create spaces with homology involving other groups. For example, if you want $H_1(K) = \mathbb{Z}/3 = [[a : 3a = 0]]$, then there must be a 1-cycle a (so that $Z_1(K) \neq 0$), and thus $\partial(a) = \emptyset = P - P$. In order that $3a \sim \emptyset$, there must be a 2-cell σ with $\partial(\sigma) = 3a$. Let K be the complex of Figure 6.30.

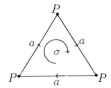

Fig. 6.30. $H_1(K) = \mathbb{Z}/3$

▷ **Exercise 6.13.** Show that the example of Figure 6.30 is not a surface.

▷ **Exercise 6.14.** Compute $H_k(K)$, $k = 0, 1, 2$, for the complexes of Figures 6.31 and 6.32.

Fig. 6.31. Weird complex

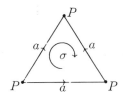

Fig. 6.32. The dunce's cap

▷ **Exercise 6.15.** Construct a complex with $H_2(K) = \mathbb{Z}\oplus\mathbb{Z}$, $H_1(K) = \mathbb{Z}/2$, $H_0(K) = \mathbb{Z}$.

6.4 Betti numbers and the euler characteristic

There is an obvious relation between the euler characteristic and the chain groups, since $C_k(K)$ is generated by the k-cells. Thus, the rank, $rk(C_k(K))$, is the number of k-cells. For a 2-complex K,

$$\chi(K) = v - e + f$$
$$= rk(C_0(K)) - rk(C_1(K)) + rk(C_2(K))$$

In general, for an n-complex K,

$$\chi(K) = rk(C_0(K)) - rk(C_1(K)) + rk(C_2(K)) - + \cdots + (-1)^n rk(C_n(K))$$

There is also a more subtle relation between the integral homology groups and the euler characteristic; to explore this we must delve a little deeper in group theory. Denote the kth boundary homomorphism on the chain groups of a complex K by $\partial_k : C_k(K) \longrightarrow C_{k-1}(K)$. Note that $Z_k(K) = ker(\partial_k)$ by Exercise 6.4 and $B_k(K) = R(\partial_{k+1})$ by Exercise 6.5. Let

$$c_k = rk(C_k(K))$$
$$z_k = rk(Z_k(K))$$
$$b_k = rk(B_k(K))$$

By Theorem A.13, for $k = 1, 2, \ldots, n$,

$$c_k = z_k + b_{k-1}$$

(6.23) Definition. *The betti numbers of a complex K are $\beta_k = rk(H_k(K))$.*

By Definition 6.14, $H_k(K) = Z_k(K)/B_k(K)$, so Theorem A.10 implies that

$$\beta_k = z_k - b_k$$

(6.24) Theorem. *Let K be an n-complex. Then*

$$\chi(K) = \beta_0 - \beta_1 + \beta_2 - + \cdots + (-1)^n \beta_n$$

Proof. Using the results above, we have

$$\beta_0 - \beta_1 + \beta_2 - + \cdots + (-1)^n \beta_n$$
$$= (z_0 - b_0) - (z_1 - b_1) + (z_2 - b_2) - + \cdots + (-1)^n (z_n - b_n)$$
$$= z_0 - (b_0 + z_1) + (b_1 + z_2) - + \cdots + (-1)^n (b_{n-1} + z_n) + (-1)^{n+1} b_n$$
$$= z_0 - c_1 + c_2 - + \cdots + (-1)^n c_n + (-1)^{n+1} b_n$$

For any 0-chain C, $\partial(C) = \emptyset$ so $Z_0(K) = C_0(K)$ and, thus, $z_0 = c_0$. The group $B_n(K) = 0$ for a n-complex K because no n-chain except $\emptyset$ can be the boundary of anything since there are no $(n+1)$-cells, so $b_n = 0$.

$$\beta_0 - \beta_1 + \beta_2 - + \cdots + (-1)^n \beta_n = c_0 - c_1 + c_2 - + \cdots + (-1)^n c_n + 0$$
$$= c_0 - c_1 + c_2 - + \cdots + (-1)^n c_n$$
$$= \chi(K)$$

$\square$

The betti numbers are topological invariants containing a lot of information. They can be combined as above to form the euler characteristic. From Theorem 6.20, note that if S is an orientable surface, $\beta_2 = 1$, and if S is non-orientable, then $\beta_2 = 0$. Thus, β_2 determines the orientability of the surface. The number of connected components in the space is obviously given by β_0.

▷ **Exercise 6.16.** Reclassify the surfaces of Exercises 4.16 and 4.17 as efficiently as you can.

▷ **Exercise 6.17.** Show that the n-dimensional ball $\mathbb{B}^n$ has $\chi(\mathbb{B}^n) = 1$.

▷ **Exercise 6.18.** Show that

$$\chi(\mathbb{S}^n) = \begin{cases} 0, & \text{if } n \text{ is odd} \\ 2, & \text{if } n \text{ is even} \end{cases}$$

Chapter 7

Cellular functions

7.1 Cellular functions

In most fields of mathematics, we study some class of objects and also the functions appropriate to the objects of study. In linear algebra, the objects studied are vector spaces and the corresponding functions, called linear transformations and usually represented by matrices, are those which preserve the vector space structure. In geometry, we study isometries, such as rotations: functions that do not change the geometric properties of length, angle measure, area, and volume. In abstract algebra, the objects of study are groups and the appropriate functions are homomorphisms, which preserve the algebraic properties of the groups. The best possible homomorphism is an isomorphism; isomorphic groups are essentially identical algebraically.

In topology, continuous functions preserve the important topological properties. By Theorem 2.29 or 3.15, connectedness is a topological property, and by Exercise 2.28 or Theorem 3.19, compactness is another. A homeomorphism may be thought of as the best possible type of continuous function, and homeomorphic spaces are considered the same in topology. In this section we wish to investigate the appropriate functions for the study of complexes on topological spaces, especially the surfaces to which we have devoted so much time. A complex has two structures: that of the topological space underlying the complex, and the subdivision of the complex into cells. The corresponding functions will have to preserve this dual nature. In particular, it would be nice if these functions, yet to be defined, would also induce nice functions on the homology of the complex. Since homology has an algebraic group structure, we want the functions to induce homomorphisms on the homology groups. The best of all functions would be homeomorphisms on the underlying topological space and also isomorphisms on the homology groups.

After a topological space is given additional structure as a cell complex or a simplicial complex, a merely continuous function is no longer entirely adequate. For example, consider the homeomorphism f that rotates the circle 90° counterclockwise. If the circle is represented by the complex K of Figure 7.1, note that $f(Q)$, shown in Figure 7.2, lands in the middle of edge a, and $f(a)$ consists of half the a edge, the vertex P, and half the b edge. Thus, f does not preserve the cell structure of the complex.

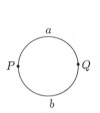

Fig. 7.1. The complex K

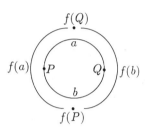

Fig. 7.2. $f(K)$

One way of dealing with this difficulty is to modify the complex so that it respects the cell structure. If the circle is represented by the complex K' of Figure 7.3, then for each cell $\sigma \in K'$, the same function f gives $f(\sigma)$ is a cell in K' as shown in Figure 7.4.

Fig. 7.3. The complex K'

Fig. 7.4. $f(K')$

The function f of the preceding examples is a homeomorphism on the underlying space (the circle), but may or may not treat the complex structure of the space nicely. In order for the function to act nicely in conjunction with the homology groups, we need to be sure that f preserves the cells of the complex.

Recall that we have yet to prove that the homology groups do not depend on the particular complex chosen to represent a space, so that must be one of our goals. As an example of other types of information that we can hope to derive from this study, recall the applications of Section 2.6. Upon careful inspection, it appears that the key argument in all of those theorems was the construction of a function from a connected space to a non-connected space, implying a contradiction and inducing the desired result. At that time, connectedness was one of the few topological properties we knew. Knowing so many more now, we anticipate similar results in later chapters.

(7.1(a)) Definition. *Let K and L be complexes and $f : |K| \longrightarrow |L|$ a continuous function. Then $f : K \longrightarrow L$ is a cellular function if*

(1) *for each cell $\sigma \in K$, $f(\sigma)$ is a cell in L;*
(2) *$dim(f(\sigma)) \leq dim(\sigma)$.*

The complexes K and L in Definition 7.1(a) can also be directed, if condition (1) is modified. It is not necessary that $f(\sigma)$ have the orientation inherited from σ, only that the cells coincide.

(7.1(b)) Definition. *Let K and L be directed complexes and $f : |K| \longrightarrow |L|$ a continuous function. Then $f : K \longrightarrow L$ is a cellular function if*

(1) *for each directed cell $\sigma \in K$, $f(\sigma) = \pm\tau$ where τ is a directed cell in L;*
(2) *$dim(f(\sigma)) \leq dim(\sigma)$.*

For any (directed or not directed) complex K and a vertex P, a continuous function $f : |K| \to P$ can be defined by $f(x) = P$ for all $x \in |K|$. This function is always cellular.

An example of a function that is not cellular is the continuous function $g : |K'| \longrightarrow |L'|$ defined by letting g squash the triangle flat as in Figure 7.5.

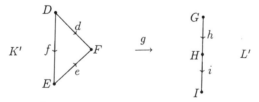

Fig. 7.5. g is not a cellular function

One way of describing g is to list what it does to each cell:

$$g(D) = G, \qquad g(d) = h$$
$$g(E) = I, \qquad g(e) = -i$$
$$g(F) = H, \qquad g(f) = h + i$$

Note that $g(f)$ is not a cell, but a union of two cells. However, we can slightly modify g to give g', which is cellular; see Figure 7.6.

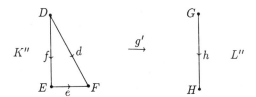

Fig. 7.6. g' is a cellular function

There are two ways of dealing with continuous functions which are not cellular: either modify the complex structure or change the function.

On the other hand, there are functions which respect the cellular structure but are not continuous on the underlying topological spaces. Consider the standard representations of the torus and Klein bottle below, and define a function f on the cells of $\mathbb{T}^2$ as in Figure 7.7.

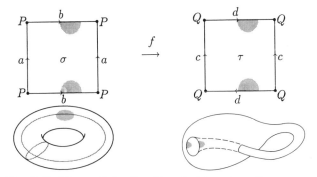

Fig. 7.7. f is a cellular function, but is not continuous

$$f(\sigma) = \tau, \qquad f(P) = Q$$
$$f(a) = c, \qquad f(b) = d$$

That f is not continuous may be seen by considering the image of the neighborhood pictured in Figure 7.7. In $\mathbb{T}^2$, the half-discs are glued to form a connected disc, but the image of this disc in $\mathbb{K}^2$ is disconnected. This contradicts Theorem 2.29, so f is not continuous. Such functions are not permitted in Definition 7.1.

We wish to compare what a cellular function $f : K \longrightarrow L$ does on the underlying topological spaces, $|K|$ and $|L|$, and what it does algebraically to the homology groups, $H_k(K)$ and $H_k(L)$. Since the homology groups are defined in terms of the chain groups $C_k(K)$ and the boundary function $\partial : C_k(K) \longrightarrow C_{k-1}(K)$, we must investigate the interaction of the cellular function f, the chain groups C_k, and the boundary homomorphism ∂. Recall that the groups $Z_k(K)$ and $B_k(K)$ were defined by ∂. We need a homomorphism $f_k : C_k(K) \longrightarrow C_k(L)$ which reflects algebraically the action of f on the k-cells of K and L.

(7.2) Definition. *Let $f : K \longrightarrow L$ be a cellular map. Define*

$$f_k : C_k(K) \longrightarrow C_k(L)$$

for each k-cell $\sigma \in K$ by

$$f_k(\sigma) = \begin{cases} f(\sigma), & \text{if } f(\sigma) \text{ is a } k\text{-cell in } L \\ \emptyset, & \text{if } \dim(f(\sigma)) < k \end{cases}$$

For a k-chain $C = a_1\sigma_1 + a_2\sigma_2 + \cdots + a_n\sigma_n \in C_k(K)$ where $\sigma_1, \sigma_2, \ldots, \sigma_n$ are k-cells in K and $a_1, a_2, \ldots, a_n \in \mathbb{Z}$ define

$$f_k(C) = f_k(a_1\sigma_1 + a_2\sigma_2 + \cdots + a_n\sigma_n)$$
$$= a_1 f_k(\sigma_1) + a_2 f_k(\sigma_2) + \cdots + a_n f_k(\sigma_n)$$

The cellular function may collapse a k-cell $\sigma \in K$ onto a lower-dimensional cell in L, but $C_k(L)$ contains only k-dimensional chains. Thus, f_k must be instructed to ignore lower-dimensional cells, so we define $f_k(\sigma) = \emptyset$ whenever $\dim(f(\sigma)) < \dim(\sigma)$.

(7.3) Theorem. *If $f : K \longrightarrow L$ is a cellular function and the function $f_k : C_k(K) \longrightarrow C_k(L)$ is induced by f, then f_k is a homomorphism.*

Proof. We must show that for any k-chains $C, D \in C_k(K)$,

$$f_k(C + D) = f_k(C) + f_k(D)$$

Since C and D are k-chains in K, they may be written as

$$C = a_1\sigma_1 + a_2\sigma_2 + \cdots + a_n\sigma_n$$
$$D = b_1\sigma_1 + b_2\sigma_2 + \cdots + b_n\sigma_n$$

where $\sigma_1, \sigma_2, \ldots, \sigma_n$ are k-cells in K. It follows from Definition 7.2 that

$$\begin{aligned}
f_k(C+D) &= f_k((a_1+b_1)\sigma_1 + (a_2+b_2)\sigma_2 + \cdots + (a_n+b_n)\sigma_n) \\
&= (a_1+b_1)f_k(\sigma_1) + (a_2+b_2)f_k(\sigma_2) + \cdots + (a_n+b_n)f_k(\sigma_n) \\
&= [a_1 f_k(\sigma_1) + \cdots + a_n f_k(\sigma_n)] + [b_1 f_k(\sigma_1) + \cdots + b_n f_k(\sigma_n)] \\
&= f_k(a_1\sigma_2 + \cdots + a_n\sigma_n) + f_k(b_1\sigma_2 + \cdots + b_n\sigma_n) \\
&= f_k(C) + f_k(D)
\end{aligned}$$

$\square$

(7.4) Example. Let K be the complex on the triangle, L the line segment, and f the function which squashes the triangle flat as in Figure 7.8.

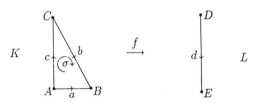

Fig. 7.8. $f : K \longrightarrow L$

The chain groups of K and L are given by

$$\begin{array}{ll}
C_2(K) = [[\sigma]] \simeq \mathbb{Z}, & C_2(L) = \{\emptyset\} \simeq 0 \\
C_1(K) = [[a,b,c]] \simeq \mathbb{Z} \oplus \mathbb{Z} \oplus \mathbb{Z}, & C_1(L) = [[d]] \simeq \mathbb{Z} \\
C_0(K) = [[A,B,C]] \simeq \mathbb{Z} \oplus \mathbb{Z} \oplus \mathbb{Z}, & C_0(L) = [[D,E]] \simeq \mathbb{Z} \oplus \mathbb{Z}
\end{array}$$

Definition 7.2 says that if we can figure out what the continuous function f does for each generating cell, then f_k is naturally defined on chains by taking combinations of these cells. Take one dimension at a time.

There is only one 2-cell, σ, in K, and $f(\sigma) = d$. Since $\dim(d) < \dim(\sigma)$, we define $f_2(\sigma) = \emptyset$. Thus, for all integer multiples $n \cdot \sigma$, $f_2(n \cdot \sigma) = \emptyset$. The homomorphism $f_2 : C_2(K) \simeq \mathbb{Z} \longrightarrow C_2(L) \simeq 0$ is the zero homomorphism, which takes each element of $\mathbb{Z}$ to 0.

On the 1-chains, $f(a) = E$, $f(b) = d$ and $f(c) = -d$, so $f_1(a) = \emptyset$, $f_1(b) = d$, $f_1(c) = -d$. Any 1-chain in K can be written in the form $\ell a + mb + nc$ for $\ell, m, n \in \mathbb{Z}$, and so

$$f_1(\ell a + mb + nc) = \ell f_1(a) + m f_1(b) + n f_1(c)$$
$$= \ell(\emptyset) + m(d) + n(-d)$$
$$= (m - n)d$$

The homomorphism $f_1 : C_1(K) \simeq \mathbb{Z} \oplus \mathbb{Z} \oplus \mathbb{Z} \longrightarrow C_1(L) \simeq \mathbb{Z}$ takes a triple (ℓ, m, n) (which stands for $\ell a + mb + nc$) and gives $f_1(\ell, m, n) = (m-n) \in \mathbb{Z}$, where $(m - n)$ stands for $(m - n)d$.

On the 0-cells of K, $f(A) = E$, $f(B) = E$, $f(C) = D$, so $f_0(A) = E$, $f_0(B) = E$, $f_0(C) = D$. Thus, f_0 takes any 0-chain of the form $\ell A + mB + nC$ to

$$f_0(\ell A + mB + nC) = \ell f_0(A) + m f_0(B) + n f_0(C)$$
$$= \ell(E) + m(E) + n(D)$$
$$= (\ell + m)E + nD$$

The homomorphism $f_0 : C_0(K) \simeq \mathbb{Z} \oplus \mathbb{Z} \oplus \mathbb{Z} \longrightarrow C_0(L) \simeq \mathbb{Z} \oplus \mathbb{Z}$ takes a triple (ℓ, m, n) in $\mathbb{Z} \oplus \mathbb{Z} \oplus \mathbb{Z}$ (which stands for $\ell A + mB + nC$) and gives $f_0(\ell, m, n) = (\ell + m, n) \in \mathbb{Z} \oplus \mathbb{Z}$, where $(\ell + m, n)$ stands for $(\ell + m)E + nD$.

These homomorphisms can be represented by matrices:

$$f_2 = (0)$$
$$f_1 = (0 \quad 1 \quad -1)$$
$$f_0 = \begin{pmatrix} 1 & 1 & 0 \\ 0 & 0 & 1 \end{pmatrix}$$

In this form, for example, $f_0(\ell A + mB + nC)$ is computed by

$$f_0(\ell A + mB + nC) = \begin{pmatrix} 1 & 1 & 0 \\ 0 & 0 & 1 \end{pmatrix} \cdot \begin{pmatrix} \ell \\ m \\ n \end{pmatrix} = \begin{pmatrix} \ell + m \\ n \end{pmatrix} = (\ell + m)E + nD$$

The representation of the functions f_k by these matrices is clearly dependent on knowing the generators of $C_k(K)$ and $C_k(L)$ and the order in which they are listed.

▷ **Exercise 7.1.**   Find the chain groups and describe the action of the homomorphisms

$$f_k : C_k(K) \longrightarrow C_k(L)$$

$k = 0, 1, 2$, where $K = L =$ the square $ABCD$ and cellular function f rotates the square by $90°$ clockwise as in Figure 7.9.

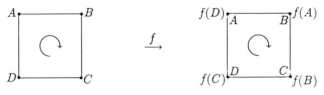

Fig. 7.9. $f : K \longrightarrow L$

▷ **Exercise 7.2.** Find the chain groups and describe the action of the homomorphisms $g_k : C_k(K') \longrightarrow C_k(L')$ where K is the cylinder, L the circle forming the lower component of the boundary of the cylinder, and the cellular function g which squashes the cylinder onto this circle as in Figure 7.10.

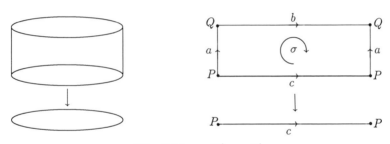

Fig. 7.10. $g : K' \longrightarrow L'$

As in both Chapters 4 and 6, it is convenient for notational reasons to sometimes deal only with simplicial or triangulated complexes. Recall that a directed n-simplex σ is completely determined by its $(n+1)$ vertices, and is written

$$\sigma = \langle v_0, v_1, v_2, \ldots, v_n \rangle$$

A cellular function of simplicial complexes is called a *simplicial function*. If a function is simplicial, then $f(\sigma)$ must be a simplex with vertices $f(v_0), f(v_1), \ldots, f(v_n)$. If f collapses σ, then some of these vertices will coincide. It, therefore, suffices to determine what the function does to each of the vertices. For non-triangulated spaces, additional information is required, since $ABCD$ can denote either a rectangle or a solid tetrahedron (recall Figure 4.33). Simplicial functions have the advantage of only needing to determine the destination of the vertices, but the disadvantage of usually requiring more cells than a cell complex. We noted in Chapter 4 that any 2-complex can be triangulated.

(7.5) Theorem. *Let K, L, and M be cell complexes, with cellular functions $f : K \longrightarrow L$ and $g : L \longrightarrow M$. Then the composition induced on the chain groups is $(g \circ f)_k = g_k \circ f_k : C_k(K) \longrightarrow C_k(M)$.*

Proof. We prove the result only for the case where K, L, and M are triangulated, and f and g are simplicial functions. We first prove the result for a single simplicial k-cell σ, and then generalize to k-chains. If σ is a k-simplex, then $f(\sigma)$ is the simplex in L determined by the vertices $f(v_0), f(v_1), \ldots, f(v_k)$. Note that these vertices will determine a simplex in L, though if some $f(v_i) = f(v_j)$, then $f(\sigma)$ will not be a k-dimensional. Applying g to the simplex $f(\sigma)$, we find that $(g \circ f)(\sigma) = g(f(\sigma))$ is the simplex in M determined by the vertices $g(f(v_0)), g(f(v_1)), \ldots, g(f(v_k))$. Thus,

$$
(g \circ f)_k(\sigma) = \begin{cases} g(f(\sigma)), & \text{if } g(f(v_i)) \neq g(f(v_j)) \; \forall i \neq j \\ \emptyset, & \text{if } \exists i \neq j \text{ such that } g(f(v_i)) = g(f(v_j)) \end{cases}
$$

On the other hand,

$$
f_k(\sigma) = \begin{cases} f(\sigma), & \text{if } f(v_i) \neq f(v_j) \; \forall i \neq j \\ \emptyset, & \text{if } \exists i \neq j \text{ such that } f(v_i) = f(v_j) \end{cases}
$$

so $g(f_k(\sigma))$ is either $g(\emptyset) = \emptyset$, or, if the $f(v_i)$'s are all distinct, then $g(f_k(\sigma))$ is the simplex with vertices $g(f(v_1)), g(f(v_2)), \ldots, g(f(v_k))$. Thus, $(g \circ f)_k(\sigma) = g_k \circ f_k(\sigma)$ for any k-simplex $\sigma \in K$. It follows immediately from Definition 7.2 that $(g \circ f)_k = (g_k) \circ (f_k)$ on k-chains. $\square$

7.2 Homology and cellular functions

In the previous section, we saw how a cellular function f induces a homomorphism on the chain groups. To see that f also induces a homomorphism on the homology groups is a bit harder. We must first investigate the way in which the function f interacts with the boundary homomorphism ∂. The proof, technical as it is, is somewhat less messy if we first triangulate the complexes K and L and assume that the function is simplicial.

(7.6(a)) Lemma. *If $f : K \longrightarrow L$ is a simplicial function, where K and L are simplicial complexes, and C is an n-chain in K, then*

$$
\partial(f_n(C)) = f_{n-1}(\partial(C))
$$

In other words, algebraically it does not matter whether one applies the function f first and then the boundary operator ∂ or vice versa. Another way of stating this relationship is by a commutative diagram.

(7.7) Definition. *Let A, B, C, D be any spaces with functions $f : A \longrightarrow B$, $g : C \longrightarrow D$, $\alpha : A \longrightarrow C$ and $\beta : B \longrightarrow D$. The diagram below is commutative if $\beta \circ f = g \circ \alpha : A \longrightarrow D$; i.e., for each $x \in A$,*

$$\beta(f(x)) = g(\alpha(x)) \in D$$

$$
\begin{array}{ccc}
A & \xrightarrow{\ f\ } & B \\
\downarrow{\scriptstyle\alpha} & & \downarrow{\scriptstyle\beta} \\
C & \xrightarrow[g]{} & D
\end{array}
$$

The diagram commutes if all paths from the upper left to the lower right give the same answer. The diagram is another way to picture the equation $\beta(f(x)) = g(\alpha(x))$. The relationship among f, g, α, and β is easier to see in the diagram than written out as an equation.

Thus, Lemma 7.6 may be restated as

(7.6(b)) Lemma. *If $f : K \longrightarrow L$ is a simplicial function, where K and L are simplicial complexes, then the diagram below commutes.*

$$
\begin{array}{ccc}
C_n(K) & \xrightarrow{\ f_n\ } & C_n(L) \\
\downarrow{\scriptstyle\partial} & & \downarrow{\scriptstyle\partial} \\
C_{n-1}(K) & \xrightarrow[f_{n-1}]{} & C_{n-1}(L)
\end{array}
$$

Proof. First consider the case when σ is a simplex:

$$\sigma = \langle v_0, v_1, v_2, \ldots, v_n \rangle$$

and

$$\partial(\sigma) = \sum_{k=0}^{n} (-1)^k \langle v_0, v_1, v_2, \ldots, \overline{v_k}, \ldots, v_n \rangle$$

We must calculate $\partial(f_n(\sigma))$ and $f_n(\partial(\sigma))$ and see if they are equal. Since f is simplicial, the simplex $f(\sigma)$ has vertices $f(v_0), f(v_1), f(v_2), \ldots, f(v_n)$. This will be an n-simplex if all the vertices remain distinct, but if $f(v_i) = f(v_j)$ for some $i \neq j$, then f collapses the cell to a lower-dimensional simplex and $f_n(\sigma)$ is defined to be $\emptyset$.

$$f_n(\sigma) = \begin{cases} \langle f(v_0), f(v_1), f(v_2), \ldots, f(v_n) \rangle & \text{if } f(v_i) \neq f(v_j) \, \forall i \neq j \\ \emptyset & \text{if } \exists i \neq j \text{ so that } f(v_i) = f(v_j) \end{cases}$$

Therefore,

$$\partial(f_n(\sigma)) = \begin{cases} \partial \langle f(v_0), f(v_1), f(v_2), \ldots, f(v_n) \rangle \\ \qquad \text{if } f(v_i) \neq f(v_j) \, \forall i \neq j \\ \emptyset \qquad \text{if } f(v_i) = f(v_j) \text{ for some } i \neq j \end{cases}$$

$$\partial(f_n(\sigma)) = \sum_{k=0}^{n} \begin{cases} (-1)^k \langle f(v_0), f(v_1), \ldots, \overline{f(v_k)}, \ldots, f(v_n) \rangle \\ \qquad \text{if } f(v_i) \neq f(v_j) \, \forall i \neq j; \, i, j \neq k \\ \emptyset \qquad \text{if } f(v_i) = f(v_j) \text{ for some } i \neq j \end{cases}$$

On the other hand,

$$f(\partial(\sigma)) = f\left(\sum_{k=0}^{n} (-1)^k \langle v_0, v_1, v_2, \ldots, \overline{v_k}, \ldots, v_n \rangle \right)$$

Therefore,

$$f_{n-1}(\partial(\sigma)) = \begin{cases} \sum_{k=0}^{n} (-1)^k \langle f(v_0), f(v_1), \ldots, \overline{f(v_k)}, \ldots, f(v_n) \rangle \\ \qquad \text{if } f(v_i) \neq f(v_j) \, \forall i \neq j; \, i, j \neq k \\ \emptyset \qquad \text{if } f(v_i) = f(v_j) \text{ for some } i \neq j \end{cases}$$

Thus, for simplices, $f_{n-1}\partial(\sigma) = \partial f_n(\sigma)$. If C is an n-chain in K, then

$$C = a_1\sigma_1 + a_2\sigma_2 + \cdots + a_m\sigma_m$$

for integers $a_1, a_2, \ldots, a_m$ and n-simplices $\sigma_1, \sigma_2, \ldots, \sigma_m \in K$.

$$\begin{aligned} \partial f_n(C) &= \partial f_n(a_1\sigma_1 + a_2\sigma_2 + \cdots + a_m\sigma_m) \\ &= \partial [a_1 f_n(\sigma_1) + a_2 f_n(\sigma_2) + \cdots + a_m f_n(\sigma_m)] \\ &= a_1 \partial(f_n(\sigma_1)) + a_2 \partial(f_n(\sigma_2)) + \cdots + a_m \partial(f_n(\sigma_m)) \\ &= a_1 f_{n-1}(\partial(\sigma_1)) + a_2 f_{n-1}(\partial(\sigma_2)) + \cdots + a_m f_{n-1}(\partial(\sigma_m)) \\ &= f_{n-1}[a_1 \partial(\sigma_1) + a_2 \partial(\sigma_2) + \cdots + a_m \partial(\sigma_m)] \\ &= f_{n-1}(\partial(a_1\sigma_1 + a_2\sigma_2 + \cdots + a_m\sigma_m)) \\ &= f_{n-1}(\partial(C)) \end{aligned}$$

$\square$

We have now completed all the preliminaries needed to prove that f induces a homomorphism on the homology groups. Recall that the kth homology group can be written as $H_k(K) = Z_k(K)/B_k(K)$ and these subgroups satisfy $B_k(K) \subseteq Z_k(K) \subseteq C_k(K)$. The cycle group $Z_k(K)$ is defined as the set of k-chains $C \in C_k(K)$ such that $\partial(C) = \emptyset$. Two k-cycles C and D are homologous in $H_k(K)$ if $C \sim D$, or, equivalently, if there is some $(k+1)$-chain $E \in C_{k+1}(K)$ with $C - D = \partial(E)$. In order for f to induce a homomorphism on the homology groups, three things must be proven:

(1) A simplicial function f takes the k-cycles in K to k-cycles in L. If so, the function f will induce a function on the cycle groups: $f_k : Z_k(K) \longrightarrow Z_k(L)$. To prove this, we must show that whenever C is a k-chain in K with $\partial(C) = \emptyset$, then $\partial(f_k(C)) = \emptyset \in C_k(L)$.

(2) If $C, D \in Z_k(K)$ with $C \sim D$, then $f_k(C) \sim f_k(D)$ in $Z_k(L)$. That is, whenever $C - D = \partial(E)$ for some $E \in C_{k+1}(K)$, then $f_k(C) - f_k(D) = \partial(F)$ for some $F \in C_{k+1}(L)$.

(3) If (1) and (2) are true, then there is a well-defined function $f_k : H_k(K) \longrightarrow H_k(L)$. We must show that it is a homomorphism.

We seem to be denoting a lot of different functions by f_k, but try not to let this bother you. The only one we are really interested in is the one on H_k and the others are temporary. All of these are naturally induced by the function f, and the context makes it clear which is intended.

(7.8) Theorem. *If $f : K \longrightarrow L$ is a simplicial function, where K and L are simplicial complexes, then f induces homomorphisms on the homology groups $f_k : H_k(K) \longrightarrow H_k(L)$.*

Proof. We follow the outline above.

Step 1: If C is a k-cycle in K so $\partial(C) = \emptyset$, then by Lemma 7.6 we have

$$\partial(f_k(C)) = f_{k-1}(\partial(C))$$
$$= f_{k-1}(\emptyset)$$
$$= \emptyset$$

Thus, $f_k(Z_k(K)) \subseteq Z_k(L)$.

Step 2: If $C \sim D$ in K, then $C - D = \partial(E)$ for some $(k+1)$-chain E in K. It follows from Lemma 7.6 and Theorem 7.3 that

$$f_k(C - D) = f_k(\partial(E))$$
$$f_k(C) - f_k(D) = \partial(f_{k+1}(E))$$

Thus, $f_k(C) - f_k(D)$ is a boundary, so $f_k(C) \sim f_k(D)$ in L. Therefore, f induces a function

$$f_k : H_k(K) \longrightarrow H_k(L)$$

Step 3: That f_k is a homomorphism follows immediately from Theorem 7.3. $\square$

(7.9) Corollary. *Let K, L, and M be simplicial complexes, with simplicial functions $f : K \longrightarrow L$ and $g : L \longrightarrow M$. Then the composition of these functions induces $(g \circ f)_k = g_k \circ f_k : H_k(K) \longrightarrow H_k(M)$.*

Proof. The proof follows immediately from Theorems 7.5 and 7.8.   □

(7.10) Corollary. *If $f : K \longrightarrow L$ is a simplicial function which is a homeomorphism on the underlying spaces, then $f_k : H_k(K) \longrightarrow H_k(L)$ is an isomorphism.*

▷ **Exercise 7.3.** Prove Corollary 7.10.

Unfortunately, the converse of Corollary 7.10 is not true, as we will see in Example 7.12.

The key to Theorem 7.8 is Lemma 7.6. A slight generation of the idea of cellular functions gives:

(7.11) Definition. *Let $f : |K| \longrightarrow |L|$ be a continuous function. If, for each k-chain C in K, $f(C)$ is a k-chain in L and if the diagram*

$$
\begin{array}{ccc}
C_k(K) & \xrightarrow{\ f\ } & C_k(L) \\
\partial \downarrow & & \downarrow \partial \\
C_{k-1}(K) & \xrightarrow[\ f\ ]{} & C_{k-1}(L)
\end{array}
$$

commutes, then $f : K \longrightarrow L$ is a chain function from K to L.

In other words, a chain function need not be cellular but is one where Lemma 7.6 is defined to be true. The proof of Theorem 7.8 then shows that f induces homomorphisms $f_k : H_k(K) \longrightarrow H_k(L)$. An example of a chain function that is not cellular is the function g pictured in Figure 7.5. This function is not cellular since $g(f) = h + i$, but g is a chain function since

$$
\begin{aligned}
g(\partial(f)) &= g(D - E) \\
&= g(D) - g(E) \\
&= G - I \\
\partial(g(f)) &= \partial(h + i) \\
&= \partial(h) + \partial(i) \\
&= (H - I) + (G - H) \\
&= G - I
\end{aligned}
$$

Thus, g will induce a nice homomorphism on the homology groups, which is enough for our purposes. Functions that are not quite cellular are often chain functions, and Theorem 7.8 and its corollaries are true for them. In particular, if $f : K \longrightarrow L$ is a cellular function and K and L are triangulable cell complexes, then note that for a k-cell $\tau \in K$, this cell can be triangulated, so $\tau = a_1\sigma_1 + a_2\sigma_2 + \cdots + a_m\sigma_m$ for some choice of $a_1, a_2, \ldots, a_m$

where $a_i = \pm 1$, depending on the orientation of τ and σ_i. Thus, τ may be regarded as a k-chain in the triangulation of K. Similarly, $f(\tau)$ will be a k-chain in the triangulation of L. Such functions are easily seen to be chain functions, and so Theorem 7.8 and its corollaries are true for all cellular functions, not only simplicial ones.

7.3 Examples

We frequently use the notation $H_*(K)$ to denote the system of homology groups $H_k(K)$ for $0 \le k \le \dim(K)$, and $f_* : H_*(K) \longrightarrow H_*(L)$ is shorthand for the collection of homomorphisms of Theorem 7.8 more properly denoted by $f_k : H_k(K) \longrightarrow H_k(L)$ for $k = 0, 1, \dots, \dim(K)$.

In practice, when computing $f_* : H_*(K) \longrightarrow H_*(L)$ for a cellular or chain function f, it suffices to determine what happens to the generators of $H_*(K)$. If $H_k(K)$ is generated by $x_1, x_2, \dots, x_n$, then any $x \in H_k(K)$ can be written as

$$x = a_1 x_1 + a_2 x_2 + \cdots + a_n x_n$$

for some choice of $a_1, a_2, \dots, a_n \in \mathbb{Z}$. Since f_k is a homomorphism,

$$
\begin{aligned}
f_k(x) &= f_k(a_1 x_1 + a_2 x_2 + \cdots + a_n x_n) \\
&= a_1 f_k(x_1) + a_2 f_k(x_2) + \cdots + a_n f_k(x_n)
\end{aligned}
$$

Below, we compute f_* for some examples, by first computing the homology groups $H_*(K)$ and $H_*(L)$, and then verifying that f is cellular and finding the value of f_k on the generators of $H_k(K)$.

(7.12) Example. Let g be the function that shrinks a cylinder $\mathbb{C}$ down to the circle running around the bottom of the cylinder, where $\mathbb{C}$ is represented

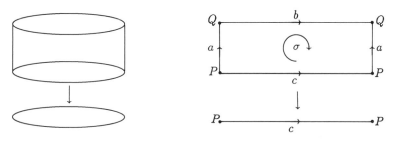

Fig. 7.11. $g : K \longrightarrow L$

by the directed complex K and the circle is represented by the subcomplex $L \subseteq K$ as in Figure 7.11.

Since $\partial(\sigma) = b - c$, there are no 2-cycles in K and $H_2(K) = 0$. Since L has no 2-cells, $H_2(L) = 0$. The function $g_2 : H_2(K) = 0 \longrightarrow H_2(L) = 0$ must be the identity or zero homomorphism defined by $g_2(0) = 1_0(0) = 0$.

The 1-cycles of K are b, c, and all combinations of these. Since, in K, $\partial(\sigma) = b - c$, we have $b \sim c$. Thus, $H_1(K) = [[b]] \simeq \mathbb{Z}$. In L, the edge c is a 1-cycle, so $H_1(L) = [[c]] \simeq \mathbb{Z}$. Since b generates $H_1(K)$ and $g(b) = c$, the homomorphism $g_1 : H_1(K) = [[b]] \simeq \mathbb{Z} \longrightarrow H_1(L) = [[c]] \simeq \mathbb{Z}$ is defined by $g_1(b) = c$, and so $g_1(n \cdot b) = n \cdot c$ for any $n \in \mathbb{Z}$. As a homomorphism $g_1 : \mathbb{Z} \longrightarrow \mathbb{Z}$ is the identity homomorphism $1_\mathbb{Z}$.

The 0-cycles of K form the group $Z_0(K) = [[P, Q]] \simeq \mathbb{Z} \oplus \mathbb{Z}$. Since $\partial(a) = Q - P$, $P \sim Q$, and $H_0(K) = [[P]] \simeq \mathbb{Z}$. The only vertex in L is P, so $H_0(L) = [[P]] \simeq \mathbb{Z}$. Since $g(P) = P$, $g_0 : H_0(K) \simeq \mathbb{Z} \longrightarrow H_0(L) \simeq \mathbb{Z}$ is also the identity homomorphism.

The system of homomorphisms g_* is thus

$$H_2 : \quad g_2 : 0 \xrightarrow{1} 0$$

$$H_1 : \quad g_1 : \mathbb{Z} \xrightarrow{1} \mathbb{Z}$$

$$H_0 : \quad g_0 : \mathbb{Z} \xrightarrow{1} \mathbb{Z}$$

or this can all be expressed by

$$H_* : \quad g_* = 1_{H_*}$$

Notice that, algebraically, $g_*(x) = x$ on each dimension; the homomorphism does not change anything, in spite of the fact that g moves lots of points. Homology, for some reason, does not see the 2-dimensionality of the cylinder. We will explain precisely why this happens in Section 9.1. The computations above show that the function g makes no change on the 1-dimensional level: g preserves the roundness of the cylinder. On the zero-dimensional level, where $H_0 = \mathbb{Z}$ tells us that the spaces are connected, g takes the connected cylinder to another connected space, the circle.

(7.13) Example. Think about what happens when one twists a circular rubber band around one's finger four times. The result winds the original circle four times around a smaller circle. Thus, each quarter of the original circle gets wrapped once around the resulting circle. Choose a complex K on the first circle which divides it in fourths, and use the standard complex L for the second circle, as in Figure 7.12.

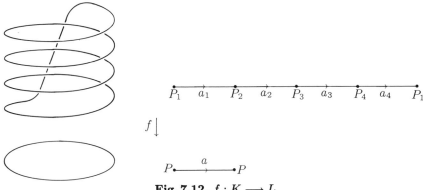

$P_1 \xrightarrow{a_1} P_2 \xrightarrow{a_2} P_3 \xrightarrow{a_3} P_4 \xrightarrow{a_4} P_1$

$f \downarrow$

$P \xrightarrow{a} P$

Fig. 7.12. $f : K \longrightarrow L$

In K, none of a_1, a_2, a_3, and a_4 are 1-cycles but their sum is a 1-cycle, since $\partial(a_1 + a_2 + a_3 + a_4) = P - P = \emptyset$. There are no 2-cells in K, so $H_1(K) = Z_1(K) = [[a_1 + a_2 + a_3 + a_4]] \simeq \mathbb{Z}$. In L, the edge b is a 1-cycle, so $H_1(K) = [[b]] \simeq \mathbb{Z}$. The function f has $f(a_1) = f(a_2) = f(a_3) = f(a_4) = b$, so $f(a_1 + a_2 + a_3 + a_4) = 4b$. Thus, f_1 takes the generator of $H_1(K)$ to four times the generator of $H_1(L)$. Thus, $f_1 : H_1(K) \simeq \mathbb{Z} \longrightarrow H_1(L) \simeq \mathbb{Z}$ is multiplication by 4. We represent this homomorphism by $\times 4$.

Since $\partial(a_1) = P_2 - P_1$, $\partial(a_2) = P_3 - P_2$, $\partial(a_3) = P_4 - P_3$, and $\partial(a_4) = P_1 - P_4$, so $P_1 \sim P_2 \sim P_3 \sim P_4$. Therefore, $H_0(K) = [[P_1]] \simeq \mathbb{Z}$. There is only one 0-cycle Q in L, so $H_0(L) = [[Q]] \simeq \mathbb{Z}$. Since $f(P_1) = Q$, f_0 takes the generator of $H_0(K)$ to the generator of $H_0(L)$, so

$$f_0 = 1_{\mathbb{Z}} : H_0(K) \simeq \mathbb{Z} \longrightarrow H_0(L) \simeq \mathbb{Z}$$

Thus, the homomorphisms f_* are defined for this example by

$$H_1 : \quad f_1 : \mathbb{Z} \xrightarrow{\times 4} \mathbb{Z}$$
$$H_0 : \quad f_0 =: \mathbb{Z} \xrightarrow{1} \mathbb{Z}$$

These can be interpreted by recalling that H_1 counts loops. Thus, f_1 says that that the generating loop in K corresponds to four times the generating loop in L. This echoes precisely the original geometrical definition of the continuous function f. On H_0, f_0 merely says that a connected space is taken to a connected space.

(7.14) Example. Take a torus and strangle it, as one does making balloon animals. This function is illustrated in Figure 7.13.

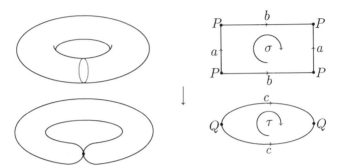

Fig. 7.13. f is a cellular function from the torus to the strangled torus

In the complex K, the only 2-cell is σ and $\partial(\sigma) = a + b - a - b = \emptyset$, and, thus, $H_2(K) = [[\sigma]] \simeq \mathbb{Z}$. In L, the 2-chains are generated by τ and $\partial(\tau) = c - c = \emptyset$, so $H_2(L) = [[\tau]] \simeq \mathbb{Z}$. Since $f(\sigma) = \tau$, the homomorphism $f_2 : H_2(K) \simeq \mathbb{Z} \longrightarrow H_2(L) \simeq \mathbb{Z}$ is the identity $1_\mathbb{Z}$.

The 1-cycles of K are generated by the loops a and b, and, thus, $H_1(K) = [[a, b]] \simeq \mathbb{Z} \oplus \mathbb{Z}$. Similarly, $H_1(L) = [[c]] \simeq \mathbb{Z}$. Since $f(a) = Q$ and $\dim(Q) < 1$, we have $f_1(a) = \emptyset$, but $f_1(b) = c$. Therefore, if one represents elements of $\mathbb{Z} \oplus \mathbb{Z}$ by $(n, m) = na + mb$, then

$$f_1(n, m) = f_1(na + mb) = nf_1(a) + mf_1(b) = 0 + m(c) = mc$$

In other words, f_1 kills the first factor and is the identity on the second. This homomorphism can be represented in matrix form by $[0 \quad 1]$.

Since $H_0(K) = [[P]] \simeq \mathbb{Z}$ and $H_0(L) = [[Q]] \simeq \mathbb{Z}$, and $f(P) = Q$, it follows that $f_0 = 1_\mathbb{Z}$.

In summary,

$$H_2 : \quad f_2 : \mathbb{Z} \xrightarrow{1} \mathbb{Z}$$
$$H_1 : \quad f_1 : \mathbb{Z} \oplus \mathbb{Z} \xrightarrow{[0 \quad 1]} \mathbb{Z}$$
$$H_0 : \quad f_0 =: \mathbb{Z} \xrightarrow{1} \mathbb{Z}$$

This can be interpreted as the function f takes the cavity on the torus to the cavity in the strangled torus, squashes the a-direction loop to a point but preserves the b-direction loop, and takes the connected space to another connected space.

Do not get discouraged. Remember that there just are not very many different homomorphisms of finitely generated abelian groups.

● *In Exercises 7.4 through 7.7, choose complexes K and L so that the function f is cellular, find the homology groups $H_*(K)$ and $H_*(L)$, describe the homomorphisms f_*, and interpret your results geometrically.*

▷ **Exercise 7.4.** Let M^2 be the Möbius band. The meridian circle of M is the circle running down the middle of the band. Let $f : M \longrightarrow \mathbb{S}^1$ squash M onto the meridian circle, where M and $\mathbb{S}^1$ are represented by the complexes K and L as in Figure 7.14.

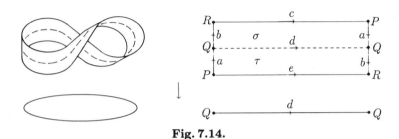

Fig. 7.14.

▷ **Exercise 7.5.** Note that the boundary of a Möbius band M is a circle. Let $f : \mathbb{S}^1 \longrightarrow M$ be the function that wraps a circle $\mathbb{S}^1$ onto this boundary; see Figure 7.15.

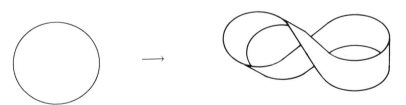

Fig. 7.15.

▷ **Exercise 7.6.** The function $f : \mathbb{R} \longrightarrow \mathbb{S}^1$ defined by
$$f(x) = (\cos(2\pi x), \sin(2\pi x))$$

wraps the real line around the circle. Choose complexes K and L to represent $\mathbb{R}$ and $\mathbb{S}^1$ so that f is a cellular function.

▷ **Exercise 7.7.** The projective plane $\mathbb{P}^2$ contains a Möbius band M. Let $f : M \longrightarrow \mathbb{P}^2$ be this inclusion.

7.4 Covering spaces

(7.15) Definition. *Let $\tilde{X}$ and X be n-dimensional manifolds where X is connected and let $p : \tilde{X} \longrightarrow X$ be a continuous function. Then $p : \tilde{X} \longrightarrow X$ is a covering space of X if every point $x \in X$ has a connected neighborhood U such that $p^{-1}(U)$ consists of a number of disjoint sets, and if V is a connected component of $p^{-1}(U)$, then $p : V \longrightarrow U$ is a homeomorphism. Neighborhoods such as U are called basic neighborhoods. The space $\tilde{X}$ is called the cover, whereas X is called the base space, and the function p is the covering function. The number of points in the set $p^{-1}(x)$ for $x \in X$ is the number of sheets in the cover.*

In other words, the function p is a *local homeomorphism*. An example of a local homeomorphism which is not a cover is the function g from the open spiral to the circle $\mathbb{S}^1$ in Figure 7.16. Although this function will take a small neighborhood on the spiral homeomorphically to a neighborhood on $\mathbb{S}^1$, note that the point **x** has no neighborhood satisfying Definition 7.15.

Fig. 7.16. g is not a covering projection

In this text, we will only consider covering spaces where both $\tilde{X}$ and X are triangulable complexes with $p : \tilde{X} \longrightarrow X$ a cellular function. Note that if $p : \tilde{X} \longrightarrow X$ is a covering space, then X has the quotient topology of Definition 3.33 with respect to the function p. The number of points in the set $p^{-1}(x)$ is the same for each point $x \in X$, so the number of sheets in a cover is well-defined, but this we do not prove.

▷ **Exercise 7.8.**   Prove that if $p : \tilde{X} \longrightarrow X$ is a covering space, then $\tilde{X}$ is an n-manifold if and only if X is an n-manifold.

Figure 7.17 is a four-sheeted covering space since every point of the base circle $\mathbb{S}^1$ has a neighborhood which looks like a bent interval. The inverse image of this neighborhood will be a collection of four open intervals in the covering circle which will be disjoint if the neighborhood on the base circle is chosen small enough, and these intervals are taken homeomorphically to the neighborhood on the base circle.

Fig. 7.17. A four-sheeted cover of the circle: $p : \mathbb{S}^1 \longrightarrow \mathbb{S}^1$

A different four-sheeted cover of the circle is illustrated in Figure 7.18.

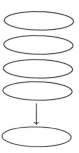

Fig. 7.18. A four-sheeted cover of the circle by four disjoint circles

For any space X, a disconnected n-fold cover can be constructed from n disjoint copies of X, but these are rarely of interest. The function of Exercise 7.6 is an infinite sheeted cover of the line over the circle.

(7.16) Example. Consider what happens to Example 7.13 if, instead of a rubber band, one uses a thin tube or torus, wrapped around itself twice instead of four times (mostly because it will be easier to draw). We represent the covering torus, $\mathbb{T}^2$, by the complex K, and the base torus by L. If L is the standard complex on the torus, then in order for the function h to be cellular, the complex K must divide the diagram for the torus in halves, as in Figure 7.19.

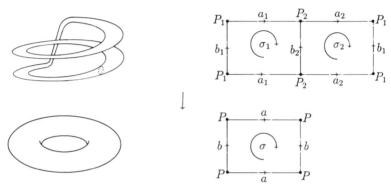

Fig. 7.19. $f : K \longrightarrow L$

In K, $\partial(\sigma_i) \neq 0$, so neither σ_1 nor σ_2 is a 2-cycle but $\sigma_1 + \sigma_2$ is, so $H_2(K) = [[\sigma_1 + \sigma_2]] \simeq \mathbb{Z}$. In L, $\partial(\sigma) = 0$, so $H_2(L) = [[\sigma]] \simeq \mathbb{Z}$. Note that $h(\sigma_i) = \sigma$ (by a fortuitous choice of notation), so for the generator of $H_2(K)$, $h_2(\sigma_1 + \sigma_2) = 2\sigma$. Thus, h_2 takes the generator of $H_2(K)$ to two times the generator of $H_2(L)$. Therefore, h_2 acts on $\mathbb{Z}$ like multiplication by 2, which we represent as $\times 2$.

The 1-cycles of K are $a_1 + a_2$, b_1, and b_2. Taking boundaries of the 2-cells, $\partial(\sigma_1) = a_1 + b_1 - a_1 - b_2 = b_1 - b_2$ and $\partial(\sigma_2) = a_2 + b_2 - a_2 - b_1 = b_2 - b_1$, so $b_1 \sim b_2$. Thus, $H_1(K) = [[a_1 + a_2, b_1]] \simeq \mathbb{Z} \oplus \mathbb{Z}$. The 1-cycles of L are a and b, so $H_1(L) = [[a, b]] \simeq \mathbb{Z} \oplus \mathbb{Z}$. As above, $h(a_1) = h(a_2) = a$ and $h(b_1) = b$, so for the generators of $H_1(K)$, $h_1(a_1 + a_2) = 2a$ and $h_1(b_1) = b$. Thus, $h_1 : H_1(K) \simeq \mathbb{Z} \oplus \mathbb{Z} \longrightarrow H_1(L) \simeq \mathbb{Z} \oplus \mathbb{Z}$ acts like multiplying the first generator by 2 and the second generator by 1. The homomorphism h_1 can be represented by the matrix:

$$h_1 = \begin{pmatrix} 2 & 0 \\ 0 & 1 \end{pmatrix}$$

The 0-cycles of K are P_1 and P_2. Since $\partial(a_1) = P_2 - P_1$ and $\partial(a_2) = P_1 - P_2$, we have $P_1 \sim P_2$, so $H_0(K) = [[P_1]] \simeq \mathbb{Z}$. The only 0-cycle in L is P, so $H_0(K) = [[P]] \simeq \mathbb{Z}$. For these 0-cycles, $h(P_1) = h(P_2) = P$, so

$h_0 : H_0(K) \simeq \mathbb{Z} \longrightarrow H_0(L) \simeq \mathbb{Z}$ is the identity homomorphism, $1_{\mathbb{Z}}$. Thus, the system of homomorphisms h_* is given by

$$H_2 : \quad h_2 : \mathbb{Z} \xrightarrow{\times 2} \mathbb{Z}$$

$$H_1 : \quad h_1 : \mathbb{Z} \oplus \mathbb{Z} \xrightarrow{\begin{pmatrix} 2 & 0 \\ 0 & 1 \end{pmatrix}} \mathbb{Z} \oplus \mathbb{Z}$$

$$H_0 : \quad h_0 =: \mathbb{Z} \xrightarrow{1} \mathbb{Z}$$

This is interpreted on H_2 as saying that the cavity in K wraps around the cavity in L twice. The torus is generated by two loops (recall from Section 3.4 that $\mathbb{T}^2 = \mathbb{S}^1 \times \mathbb{S}^1$). Then h_1 tells us that the a-direction loop (represented by $a_1 + a_2$) in K gets wrapped around twice the a-direction loop in L, but the b-direction loop only once.

(7.17) Example. Define a cellular function f on the sphere by identifying antipodal points (points directly opposite each other) on $\mathbb{S}^2$. The sphere $\mathbb{S}^2$ is represented by the complex K with σ the upper hemisphere and τ the lower. The easiest way to think about identifying antipodal points is to cut the sphere open along the equator, so that we have τ and $Int(\sigma)$, then flatten out τ and $Int(\sigma)$ into discs, rotate $Int(\sigma)$, and glue to τ, thus identifying all the points in $Int(\sigma)$ to their antipodes in $Int(\tau)$. This is illustrated in Figure 7.20. Three pairs of antipodal points are marked to help orient the sphere.

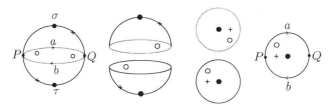

Fig. 7.20. Identifying antipodal points in $Int(\sigma)$ and $Int(\tau)$

The points on the equator must yet be identified. Identifying opposite points on the boundary circle of the disc resulting from the process of Figure 7.20 gives the projective plane $\mathbb{P}^2$. The operation described is thus a function $f : K \longrightarrow L$ where K is a complex of Figure 7.20 on the sphere $\mathbb{S}^2$ and L is a complex on the projective plane $\mathbb{P}^2$ in Figure 7.21. The complex structure K on $\mathbb{S}^2$ was chosen so that we could use the standard planar diagram for L on $\mathbb{P}^2$ and have f cellular with respect to these complexes.

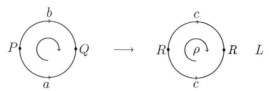

Fig. 7.21. Identifying antipodal points on the equator

In K, $\partial(\sigma) = b + a$ and $\partial(\tau) = b + a$, so $\partial(\sigma - \tau) = \emptyset$. Thus, $\sigma - \tau$ is a 2-cycle and $H_2(K) = [[(\sigma - \tau)]] \simeq \mathbb{Z}$. Since $\mathbb{P}^2$ is non-orientable, $H_2(L) = 0$. On these 2-cycles, $f(\sigma - \tau) = \rho - \rho = \emptyset$, so $f_2 : H_2(K) \simeq \mathbb{Z} \longrightarrow H_2(L) \simeq 0$ is defined by $f_2(\sigma - \tau) = 0$. Thus, f_2 is the trivial homomorphism.

Of the 1-chains of K, note that $\partial(a) = Q - P$ and also $\partial(b) = P - Q$, so $a + b$ is a 1-cycle in K and $Z_1(K) = [[a + b]]$. Since $\partial(\sigma) = a + b$, it follows that $a + b \sim 0$. Thus, $H_1(K) = 0$. In L, $\partial(c) = R - R = \emptyset$, so c is a 1-cycle. Since $\partial(\rho) = c + c = 2c$, it follows that $2c \sim 0$. The homology group is $H_1(L) = [[c : 2c = 0]] \simeq \mathbb{Z}/2$, and the induced homomorphism $f_1 : H_1(K) \simeq 0 \longrightarrow H_1(L) \simeq \mathbb{Z}/2$ is defined by $f_1(0) = 0 \in \mathbb{Z}/2$.

Since $\partial(a) = Q - P$, $Q \sim P$ in K, and thus $H_0(K) = [[P]] \simeq \mathbb{Z}$. The only 0-cycle in L is R, so $H_0(L) = [[R]] \simeq \mathbb{Z}$. Since $f(P) = R$, $f_0 : H_0(K) \simeq \mathbb{Z} \longrightarrow H_0(L) \simeq \mathbb{Z}$ is defined by $f_0(P) = R$. Therefore, f_0 is the identity homomorphism $1_{\mathbb{Z}}$.

The system of homomorphisms f_* is given by

$$H_2 : f_2 : \mathbb{Z} \xrightarrow{0} 0$$

$$H_1 : f_1 : 0 \xrightarrow{0} \mathbb{Z}/2$$

$$H_0 : f_0 : \mathbb{Z} \xrightarrow{1} \mathbb{Z}$$

This is interpreted by saying that the function f collapses the cavity in $\mathbb{S}^2$. The sphere has no non-trivial loops, so the generating loop c in $\mathbb{P}^2$ did not come from any loop on the sphere. The function f preserves the connectivity of the spaces.

(7.18) Example. The cylinder with the cellular structure and projection function $p : K \longrightarrow L$ of Figure 7.22 is a two-fold cover of the Möbius band. The covering function is defined by $p(\sigma_i) = \sigma$, $p(a_i) = a$, $p(P_i) = P$, etc., and is chosen to be cellular and to keep the notation simple.

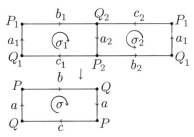

Fig. 7.22. $p : K \longrightarrow L$

▷ **Exercise 7.9.** Compute $H_*(K)$, $H_*(L)$, and the action of the homomorphisms p_* for the cover of Example 7.18.

▷ **Exercise 7.10.** Modify Example 7.18 to find a two-fold cover of the torus over the Klein bottle, and describe the action of p_* on the homology groups.

In Examples 7.17 and 7.18 and Exercise 7.10, we have constructed orientable double covers of the basic non-orientable surfaces: the Möbius band, the projective plane, and the Klein bottle. Similarly, any non-orientable surface has an orientable two-fold cover.

(7.19) Theorem. *If $\tilde{X}$ and X are surfaces with $p : \tilde{X} \longrightarrow X$ is an n-sheeted covering space, then $\chi(\tilde{X}) = n\,\chi(X)$.*

Proof. We assume without proof that the triangulation on X can be subdivided small enough that each cell σ in the subdivision will fit inside one of the basic neighborhoods of Definition 7.15. Let K be this triangulation of X and $\tilde{K}$ the corresponding triangulation on $\tilde{X}$ so that $p : \tilde{K} \longrightarrow K$ is a cellular function. Then $p^{-1}(\sigma)$ will consist of n homeomorphic copies $\sigma_1, \sigma_2, \ldots, \sigma_n \in \tilde{K}$. Thus, the number of k-cells in $\tilde{K}$ is n times the number of k-cells in K, and so $\chi(\tilde{X}) = \chi(\tilde{K}) = n\chi(K) = n\chi(X)$.   □

▷ **Exercise 7.11.**

    (1) Can the projective plane be a covering space for the Klein bottle?
    (2) What spaces can the sphere cover?
    (3) By which spaces can the two-handled torus be covered?

▷ **Exercise 7.12.** Construct an infinite cover $p : \mathbb{S}^1 \times \mathbb{R} \longrightarrow \mathbb{T}^2$ and describe the action of the homomorphisms $p_* : H_*(\mathbb{S}^1 \times \mathbb{R}) \longrightarrow H_*(\mathbb{T}^2)$. [Hint:

Consider $\mathbb{S}^1 \times \mathbb{R}$ as an infinitely long tube, and wrap this around the torus as in Exercise 7.6]

▷ **Exercise 7.13.** Draw a picture of a connected two-fold cover of the two-handled torus, $\mathbb{T}^2 \# \mathbb{T}^2$.

(7.20) Example. Here is a method of generating n-fold covers of a two-handled torus, from Massey's *A Basic Course in Algebraic Topology*. A *permutation* of the set $\{1, 2, \dots, n\}$ is a rearrangement of the set. To construct a four-fold cover of the two-handled torus, choose two permutations, P_ℓ and P_r, of the set $\{1, 2, 3, 4\}$. The notation we use is

$$P_\ell = \begin{bmatrix} 1 & 2 & 3 & 4 \\ 3 & 1 & 2 & 4 \end{bmatrix}$$

which denotes the permutation which sends 1 to 3, 2 to 1, 3 to 2, and 4 to 4. Another permutation of $\{1, 2, 3, 4\}$ is

$$P_r = \begin{bmatrix} 1 & 2 & 3 & 4 \\ 2 & 1 & 4 & 3 \end{bmatrix}$$

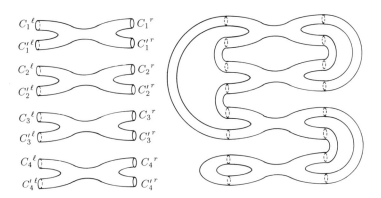

Fig. 7.23. A four-fold cover of the two-handled torus

A four-fold cover of the two-handled torus is constructed by taking four copies of the two-handled torus, and cutting each open along both of the handles, to create on each side four pairs of boundary circles which we will denote by $C^\ell{}_i$ for the ith upper circle on the left, $C'^\ell{}_i$ for the ith lower circle on the left, $C^r{}_i$ for the ith upper circle on the right, and $C'^r{}_i$ for the ith lower circle on the right, as shown in Figure 7.23. Now glue the upper and lower boundary circles on each side according to the permutations P_ℓ

and P_r given above: i.e., glue $C^{\ell}{}_1$ to $C^{\ell'}{}_3$, $C^{\ell}{}_2$ to $C^{\ell'}{}_1$, $C^{\ell}{}_3$ to $C^{\ell'}{}_2$, and $C^{\ell}{}_4$ to $C^{\ell'}{}_4$, and proceed similarly on the right side using the permutation P_r.
Note that the result is a five-handled torus, since

$$\chi(5\mathbb{T}^2) = 4\chi(2\mathbb{T}^2) = 4(-2) = -8$$

▷ **Exercise 7.14.** Draw a five-fold cover of the the two-handled torus as in Example 7.20, using the permutations

$$P_\ell = \begin{bmatrix} 1 & 2 & 3 & 4 & 5 \\ 3 & 2 & 1 & 5 & 4 \end{bmatrix}, \qquad P_r = \begin{bmatrix} 1 & 2 & 3 & 4 & 5 \\ 5 & 4 & 3 & 2 & 1 \end{bmatrix}$$

▷ **Exercise 7.15.** Adapt the method of Example 7.20 to build a three-fold cover of the Klein bottle.

Chapter 8

Invariance of homology

"A new scientific truth triumphs, not because it convinces its opponents and makes them see the light, but because the opponents eventually die, and a new generation that is familiar with it grows up."

Max Planck

8.1 Invariance of homology for surfaces

We wish to show that the homology groups of a complex K depend only on the underlying space $|K|$ and not on the particular complex chosen. We first prove this result for surfaces and indicate why the technique used in this section does not generalize to all complexes.

The proof of the invariance of the homology groups for surfaces greatly resembles the proof of Theorem 5.13 of the invariance of the euler characteristic. We assume that K is a 2-complex representing a surface, so K is composed of polygons identified along the edges and vertices. Any polygon can be subdivided into triangles, and these triangles can be further subdivided by the process of barycentric subdivision to form a triangulation of the surface. We designate this simplicial complex by K', so $|K| = |K'|$. By Theorem 4.14 (or Theorem 4.17 for surfaces with boundary) any triangulation of a surface may be reduced to a planar diagram in standard form. In order to prove that the homology groups do not depend on the complex K, we must show that the changes that turned K into the simplicial complex K', and then the changes made to K' in going through the seven steps of Theorem 4.14, do not affect the homology groups.

▷ **Exercise 8.1.** Compute the homology groups for K a triangle and for K' the barycentric subdivision of K and show that they are isomorphic.

Note that Exercise 8.1 does not imply that the homology groups of a complex are the same as the homology groups of its barycentric subdivision, since a cell cannot be viewed in isolation. As in Theorem 5.13, we first consider the effects of the three elementary types of subdivisions.

(8.1) Lemma. *Let K be a 2-complex and let L be the complex obtained by one elementary operation or its inverse on K. An elementary operation is one of the following:*

(1) *A 2-cell in K is subdivided into two 2-cells by introducing an edge between two vertices on the boundary of the original 2-cell.*

(2) *A vertex is introduced in the interior of a 2-cell in K and an edge added connecting this vertex to one of the vertices of the 2-cell.*

(3) *A 1-cell in K is subdivided into two 1-cells by introducing a new vertex in the interior of the edge.*

Then the homology groups of K and L are isomorphic.

Proof. The first type of elementary operation is pictured in Figure 8.1.

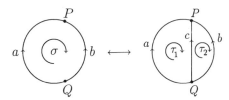

Fig. 8.1. Type 1 operation: cutting a 2-cell in two

Note that every cell of K is a cell in L with the exception of the cell σ. Define a chain function $f : K \longrightarrow L$ by

$$f(C) = \begin{cases} C & \text{if } \sigma \notin C \\ C \text{ with } \sigma \text{ replaced by } \tau_1 + \tau_2 & \text{if } \sigma \in C \end{cases}$$

where C is a k-chain in K. The function f is the identity except when the 2-cell σ is involved. To show that f is a chain function, note that

$$
\begin{aligned}
f(\partial(\sigma)) &= f(a - b) \\
&= f(a) - f(b) \\
&= a - b \\
&= (a - c) + (c - b) \\
&= \partial(\tau_1) + \partial(\tau_2) \\
&= \partial(\tau_1 + \tau_2) \\
&= \partial(f(\sigma))
\end{aligned}
$$

By Definition 7.11, f is a chain function and so induces a homomorphism on the homology groups. The homomorphism induced by f is not an isomorphism on the chain groups, since K and L have different numbers of 1- and 2-cells.

We must now show that f induces an isomorphism on the homology groups. We do this by showing that $f_k : H_k(K) \longrightarrow H_k(L)$ is 1-1 and onto for $k = 2, 1, 0$. Let $C \in H_2(K)$ such that $f_2(C) = 0$ in $H_2(L)$. Since there are no 3-cells in K or L, the only way $f_2(C)$ can be trivial is if $f_2(C)$ were trivial, i.e., $f_2(C) = \emptyset$. Thus, either $f(C) = \emptyset$, or $\dim(f(C)) < \dim(C) = 2$. However, the function f does not collapse any cells, so C must be the empty set, so C is trivial in $H_2(K)$. Thus, f_2 is 1-1. To show that f_2 is onto, let $D \in H_2(L)$. We must show that there is a chain $C \in H_2(K)$ so that $f_2(C) = D$. If D does not contain τ_1 or τ_2, then $f_2(D) = D$, so we can let $C = D$. Note that while a 2-chain might easily contain only one of τ_1 or τ_2, any 2-cycle must contain both of these since if only $\tau_1 \in D \in H_2(L)$, then $-c$ is an edge in $\partial(\tau_1) \subseteq \partial(D)$. Since D is a 2-cycle, $\partial(D) = \emptyset$. The only way $-c$ can be in $\partial(D)$ is if it is eventually canceled out by $+c$. The only other cell which has c in its boundary is τ_2, so if $\tau_1 \in D$, then $(\tau_1 + \tau_2) \in D$. Since $f(\sigma) = \tau_1 + \tau_2$, let C be the 2-chain obtained from D by replacing any occurrence of $(\tau_1 + \tau_2)$ by σ, so that $f(C) = D$ and f_2 is onto.

Let $C \in H_1(K)$ with $f_1(C) = 0$. Then $f_1(C) = \partial(D)$ for some 2-chain D in L. Since C is a 1-chain in K, C cannot involve the edge c. As above D cannot contain only one of 2-cells τ_1 and τ_2. Thus, a 2-chain D' in K is defined by D with $\tau_1 + \tau_2$ replaced by σ, and $\partial(D') = C$. Therefore, f_1 is 1-1. Let C' be a 1-cycle in $H_1(L)$. Note that since $\partial(\tau_1) = a - c$, we have $c \sim a$. Therefore, C' is homologous to a 1-chain $D' \in H_1(K)$ which does not involve the edge c. Thus, $f_1(D') = D' \sim C$, and f_1 is onto.

Since the 0-cells in K and L are the same, it is easy to see that f_0 is also an isomorphism. Thus, f induces isomorphisms on the homology groups at each level.   □

▷ **Exercise 8.2.**   Prove the other two cases of Lemma 8.1. Refer to Figures 5.26 and 5.27 for pictures of the operations.

(8.2) Theorem. *Let K be a complex and $K^{(1)}$ be the barycentric subdivision of K. Then $H_*(K) \simeq H_*(K^{(1)})$.*

Proof. This follows from Lemma 8.1 for a 2-complex K. A similar lemma can be proven for subdivisions of higher-dimensional cells, but will not be included in this text. The theorem is, thus, true for all dimensions.   □

(8.3) Theorem. *Let S represent the complex given by the standard planar diagram for a surface. If K is a 2-complex such that $|K|$ is homeomorphic to $|S|$, then $H_*(K) \simeq H_*(S)$.*

▷ **Exercise 8.3.**   Prove Theorem 8.3.

(8.4) Corollary. *If K and K' are 2-complexes with $|K| = |K'| = S$ for some surface S, then $H_*(K) \simeq H_*(K')$.*

Proof. By Theorem 8.3, $H_*(K) \simeq H_*(S)$, where $H_*(S)$ is computed from the complex given by the standard planar diagram for S. Similarly, $H_*(K') \simeq H_*(S)$, so $H_*(K) \simeq H_*(K')$.   □

The theorem and its corollary imply that the homology groups depend, for surfaces at least, only on the underlying space and not on the particular complex used to represent the space. Thus, any two 2-complexes for a surface give the same homology groups, so it makes sense to use the simplest. The key argument in the corollary is that both K and K' are subdivisions of a common planar diagram. Thus, K can be disassembled (as in Theorem 4.14) to get the planar diagram, and then the planar diagram can be subdivided to get the complex K'. The standard planar diagram forms a link between the two complexes.

The argument of Corollary 8.4 can be turned inside out, as in the following corollary:

(8.5) Corollary. *If K and K' are complexes with $|K| = |K'|$, and K and K' have a common subdivision K'', then $H_k(K) \simeq H_k(K')$.*

Proof. If K can be subdivided to obtain K'', then by repeated use of the results of Lemma 8.1, $H_k(K) \simeq H_k(K'')$. Similarly, if K' can be subdivided to obtain K'', then $H_k(K') \simeq H_k(K'')$. Thus, $H_k(K) \simeq H_k(K')$.   □

In this corollary, the complex K can be subdivided by a finite sequence of elementary subdivisions to get the complex K'', and then K'' can be disassembled using the inverses of the elementary subdivisions to get the complex K'. Thus, the existence of a common subdivision forms a link between the two complexes. For arbitrary 2-complexes (not surfaces only), one can prove the invariance of the homology groups by using Corollary 8.5 above and the 2-dimensional version of a famous topological conjecture:

(8.6) Hauptvermutung *Any two triangulations of a topological space have a common subdivision.*

This conjecture was proved for all triangulable 2-complexes in 1963 by Papakyriakopoulos, but is false in general for topological spaces of dimension greater than 2. The discovery of this failure in higher dimensions changed the direction of the study of topology away from triangulations. Alternate techniques have been developed to prove the invariance of the homology groups, which are outlined in the next section. These techniques have the added advantage of allowing the study of the action of any continuous function on the homology groups, instead of merely cellular functions.

8.2 The Simplicial Approximation Theorem

We have been quietly assuming that cellular functions are easily come by. This is actually not true: if $f : |K| \longrightarrow |L|$ is any continuous function and K and L are complexes, unless L consists of only finitely many points the probability is zero that a vertex $v \in K$ lands exactly on top of a vertex $v' \in L$ under the action of the function f. One can sometimes modify the complex L so that $f(v)$ is a vertex in L for each vertex $v \in K$, but often this is impossible.

We can modify the example of Figures 7.1 and 7.2 to let $f : \mathbb{S}^1 \longrightarrow \mathbb{S}^1$ be the function that rotates the circle by $\sqrt{2}\pi$ radians. If $f : K \longrightarrow K$ were a cellular function and $v_1 = (1, 0)$ a vertex in K, then it would follow that $v_2 = f(v_1) = (\cos\sqrt{2}\pi, \sin\sqrt{2}\pi)$ must also be a vertex in K. But then $f(v_2)$ must also be a vertex $v_3 = f(v_2) = (\cos(2\sqrt{2}\pi), \sin(2\sqrt{2}\pi))$ and then $v_4 = f(v_3) = (\cos(3\sqrt{2}\pi), \sin(3\sqrt{2}\pi))$ must be a vertex in K, etc. Since $\sqrt{2}$ is irrational, this process never ends, so K must have infinitely many vertices, which is not allowed for a complex on a compact space.

The solution is to alter the function rather than the complex. We wish to approximate the continuous function $f : |K| \longrightarrow |L|$ by a cellular function $g : K \longrightarrow L$, keeping f and g close enough to each other so that g acts almost the same as f. We must decide when two functions are close to each other. The nearest thing we have to a definition of "closeness" is being in the same neighborhood, as, for example, in the definition of a limit point (Definition 2.2). We need to define open neighborhoods which are compatible with the cell structure of a complex. As before, when facing a very technical objective, we choose to work only with simplicial complexes.

For notational convenience, for a vertex $v \in K$, define $Int(v) = v$.

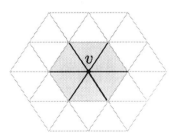

Fig. 8.2. $st_K(v)$

(8.7) Definition. *Let K be a simplicial complex with vertex v. The star of v is the set*

$$st_K(v) = \bigcup_{v < \sigma \in K} Int(\sigma).$$

The star of a vertex (see Fig. 8.2) is an open set which acts like a neighborhood of v and is also related to the simplicial structure of K. Note that $st_K(v)$ consists of whole interiors of cells. Thus, if τ is a simplex with $Int(\tau) \cap st_K(v) \neq \emptyset$, then $Int(\tau) \subseteq st_K(v)$.

▷ **Exercise 8.4.** If K is a simplicial complex, prove that $\{st_K(v) : v \in K\}$ forms a finite open cover of $|K|$. That is,

$$|K| \subseteq \bigcup_{v \in K} st_K(v).$$

▷ **Exercise 8.5.** For the complex K of Figure 8.3 describe the following:

(1) $st_K(v_i)$ for $i = 0, 1, 2, 3, 4, 5, 6$
(2) $st_K(v_1) \cap st_K(v_2) \cap st_K(v_3)$
(3) $st_K(v_4) \cap st_K(v_5)$
(4) $\bigcup_{i=0}^{6} st_K(v_i)$

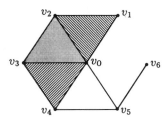

Fig. 8.3. The simplicial complex K

We wish to find some mathematical way of saying that a cellular function $g : K \longrightarrow L$ is a good approximation for a continuous function f. These functions will be close if $f(\sigma)$ is close to $g(\sigma)$ for every cell $\sigma \in K$. Note that $g(\sigma)$ will be a cell in L, but $f(\sigma)$ will not be a cell in general. Refer to Figure 8.4.

Recall that if $\sigma = \langle v_0, v_1, \ldots, v_k \rangle$ is a simplex in K and g is a simplicial function, then $g(\sigma) = \langle g(v_0), g(v_1), \ldots, g(v_k) \rangle$. Whatever $f(\sigma)$ may be, we at least know that it connects the points $f(v_0), f(v_1), \ldots, f(v_k)$. Since f is continuous, f will be reasonably close to g if $f(v)$ is close to $g(v)$ for each vertex $v \in K$.

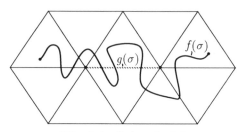

Fig. 8.4. $f(\sigma)$ and $g(\sigma)$

(8.8) Definition. *Let K and L be simplicial complexes. If $f : |K| \longrightarrow |L|$ is a continuous function, then a simplicial function $g : K \longrightarrow L$ is a simplicial approximation of f if for every vertex $v \in K$*

$$f(st_K(v)) \subseteq st_L(g(v)).$$

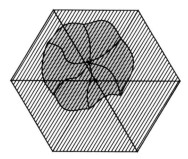

Fig. 8.5. $f(st_K(v)) \subseteq st_L(g(v))$

A cellular function g is a simplicial approximation for f if the image under f of a star neighborhood of a vertex v fits inside the star neighborhood of $g(v)$ for each vertex v, as in Figure 8.5. Thus, $f(v)$ and $g(v)$ must be close to each other. Since f is continuous, it follows that $f(\sigma)$ is is close to $g(\sigma)$ for higher-dimensional cells.

Note that not all continuous functions can be approximated by simplicial functions. For example, let $f : |K| \longrightarrow |L|$ be the function pictured in Figure 8.6, where $f(A) = A'$, $f(B) = B'$, and $f(C) = C'$, and f bends each line segment in the middle. Thus, f takes the midpoint of the edge $a = AB$ to the point F', etc.

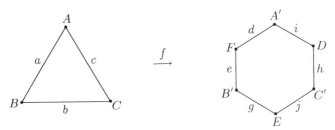

Fig. 8.6. $f : K \longrightarrow L$

The function f is clearly continuous, but is not simplicial since the image of the cell a is $f(a) = d + e$ which is not a simplex. Note that $st_K(A) = a + c$. Therefore, $f(st_K(A)) = e + d + i + h$, but $st_L(f(A)) = st_L(A') = d + i$. If we try to find a simplicial approximation g for f and let $g(A) = A'$, the only logical choice, then $f(st_K(A)) \not\subset st_L(g(A))$. Thus, f has no simplicial approximation.

However, if we are willing to take the barycentric subdivision of the complex K as many times as necessary, we can break the cells of K down until they are small enough that $f(st_K(v))$ can fit inside $st_L(g(v))$ to allow for a simplicial approximation.

▷ **Exercise 8.6.**   Take the barycentric subdivision of the complex K in Figure 8.6 to find a simplicial approximation for the function f.

We need a measure of how small the cells of a simplicial complex are.

(8.9) Definition. *If K is a simplicial complex and σ is a simplex in K, then the diameter of σ is*

$$diam(\sigma) = \max\{\|x - y\| : x, y \in \sigma\}$$

The mesh of K is

$$mesh(K) = \max\{diam(\sigma) : \sigma \in K\}$$

The diameter of a simplex is just the length of its longest side, and the mesh of a complex is the largest diameter of any of its cells. Since we only allow a finite number of cells, this leads to no difficulties.

▷ **Exercise 8.7.**   If K is a single 2-simplex represented by an equilateral triangle and $K^{(1)}$ is the first barycentric subdivision of K, show that $mesh(K^{(1)}) = \frac{2}{3}mesh(K)$.

(8.10) Lemma. *For an n-dimensional simplicial complex K, with barycentric subdivision $K^{(1)}$,*

$$mesh(K^{(1)}) \leq \frac{n}{n+1} mesh(K).$$

Proof. The proof of this lemma is omitted.

(8.11) Simplicial Approximation Theorem. *Let K and L be finite simplicial complexes with a continuous function $f : |K| \longrightarrow |L|$. Then there exists a simplicial approximation $g : K^{(k)} \longrightarrow L$ for f, where $K^{(k)}$ is the kth barycentric subdivision of K.*

Proof. We wish to find a simplicial function $g : K^{(k)} \longrightarrow L$ so that for each vertex $v \in K^{(k)}$, $f(st_{K^{(k)}}(v)) \subseteq st_L(g(v))$. Since g is to be simplicial, it suffices to define $g(v) \in L$ for each vertex $v \in K^{(k)}$. Thus, we must find a vertex $v' = g(v) \in L$ with $f(st_{K^{(k)}}(v)) \subseteq st_L(v')$. Therefore, we must show that for every $v \in K^{(k)}$ there is a $v' \in L$ such that $st_{K^{(k)}}(v) \subseteq f^{-1}(st_L(v'))$. By Lemma 8.10, $mesh(K^{(k)}) \leq (\frac{n}{n+1})^k mesh(K)$ and note that $\lim_{k \to \infty}(\frac{n}{n+1})^k = 0$. By taking the barycentric subdivision as often as necessary, the size of the cells can be made as small as we like, so eventually we can fit each one of the $st_{K^{(k)}}(v)$'s inside one of finitely many $f^{-1}(st_L(v'))$'s. In other words, for every vertex $v \in K^{(k)}$,

$$f(st_{K^{(k)}}(v)) \subseteq st_L(v') \text{ for some } v' \in L$$

Define the simplicial function $g : K^{(k)} \longrightarrow L$ on the vertices of $K^{(k)}$ by $g(v) = v'$, where v' is the vertex found above. The definition of g is extended to higher-dimensional cells by defining for a simplex $\sigma = \langle v_0, v_1, \dots, v_n \rangle$ in K, $g(\sigma) = \langle g(v_0), g(v_1), \dots, g(v_n) \rangle \in L$, so that g is simplicial. By Definition 8.8, g is a simplicial approximation for f. □

Unfortunately, g is not uniquely defined; f may have more than one simplicial approximation. However, any simplicial approximation allows us to define a function on the homology groups.

(8.12) Definition. *Let K and L be simplicial complexes with a continuous function $f : |K| \longrightarrow |L|$. A homomorphism $f_* : H_*(K) \longrightarrow H_*(L)$ is defined by $f_* = g_*$ where g is any simplicial approximation of f.*

Properly speaking, g_* is a homomorphism on $H_*(K^{(k)})$ rather than $H_*(K)$, but by Theorem 8.2 the barycentric subdivision induces an isomorphism $H_*(K) \simeq H_*(K^{(k)})$. The rest of this section is devoted to very slowly proving Corollary 8.19, which says that f_* is well-defined: i.e., if g and h are two different simplicial approximations of f, then $g_* = h_*$. If you are willing to take this fact on trust, the remainder be omitted without loss of clarity, but please note that Corollaries 8.20 and 8.21, the generalizations of Theorems 7.9 and 7.10, will be used in the following chapters.

If g and h are two different simplicial approximations for f, then it must be true that they are somehow related. We need to build a bridge between them. The appropriate bridge is called a *chain homotopy*.

If K is a simplicial complex and I the unit interval, there is a natural cell complex structure on $K \times I$, defined by letting the cells of $K \times I$ be of one of the forms $\iota \cdot \{0\}$, $\sigma \times \{1\}$, or $\sigma \times I$, where σ is a simplex in K. Note that if σ has dimension k, then $\sigma \times \{0\}$ and $\sigma \times \{1\}$ are k-cells, but $\sigma \times I$ is a $(k+1)$-cell. For example, for a vertex v, $v \times I$ is the edge or 1-cell from $v \times \{0\}$ to $v \times \{1\}$, and so $\partial(v \times I) = (v \times \{1\}) - (v \times \{0\})$. For a 1-simplex or edge $a \in K$ from vertex v_0 to v_1, $a \times I$ is the rectangle directed as in Figure 8.7. Figure 8.8 illustrates $\sigma \times I$ for a 2-simplex σ.

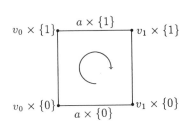

Fig. 8.7. $a \times I$

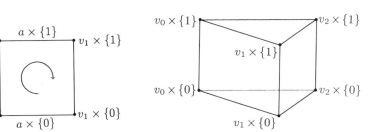

Fig. 8.8. $\sigma \times I$

$$\begin{aligned}
\partial(a \times I) &= (a \times \{1\}) - (v_1 \times I) - (a \times \{0\}) + (v_0 \times I) \\
&= (a \times \{1\}) - (a \times \{0\}) - [(v_1 - v_0) \times I] \\
&= (a \times \{1\}) - (a \times \{0\}) - (\partial(a) \times I)
\end{aligned}$$

In general, $\partial(\sigma \times I) = \sigma \times \{1\} - \sigma \times \{0\} - (\partial(\sigma) \times I)$ if one orients $\sigma \times I$ the most obvious way.

We can subdivide the cell complex on $K \times I$ so that it is simplicial. The cells of the form $\sigma \times \{0\}$ and $\sigma \times \{1\}$ are already simplices, so we need only consider cells of the form $\sigma \times I$. Since one soon tires of writing out $v_i \times \{0\}$ and $v_i \times \{1\}$, let us denote $v_i \times \{0\} = \check{v}_i$ and $v_i \times \{1\} = \hat{v}_i$. For a 1-simplex $a = \langle v_0, v_1 \rangle$ in K, $a \times I$ is a rectangle with vertices $\hat{v}_0, \hat{v}_1, \check{v}_0,$ and $\check{v}_1$. Then $a \times I = \langle \hat{v}_0, \hat{v}_1, \check{v}_1 \rangle - \langle \hat{v}_0, \check{v}_0, \check{v}_1 \rangle$. For a 2-simplex $\sigma \in K$ with $\sigma = \langle v_0, v_1, v_2 \rangle$, note that the 3-cell $\sigma \times I$ can be subdivided as in Figure 8.10, and

$$\sigma \times I = \langle \hat{v}_0, \hat{v}_1, \hat{v}_2, \check{v}_2 \rangle - \langle \hat{v}_0, \hat{v}_1, \check{v}_1, \check{v}_2 \rangle + \langle \hat{v}_0, \check{v}_0, \check{v}_1, \check{v}_2 \rangle.$$

In general, for n-simplex $\sigma = \langle v_0, v_1, \ldots, v_n \rangle$ in K,

$$\sigma \times I = \sum_{k=0}^{n} (-1)^k \langle \hat{v}_0, \hat{v}_1, \ldots, \hat{v}_k, \check{v}_k, \ldots, \check{v}_n \rangle$$

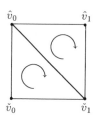

Fig. 8.9. A simplicial complex for the rectangle $a \times I$

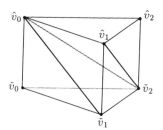

Fig. 8.10. A simplicial complex for the prism $\sigma \times I$

▷ **Exercise 8.8.**

(1) Show that $\partial(a \times I) = (a \times \{1\}) - (a \times \{0\}) - (\partial(a) \times I)$ for the triangulated 1-simplex a of Figure 8.9.

(2) Show that $\partial(\sigma \times I) = (\sigma \times \{1\}) - (\sigma \times \{0\}) - (\partial(\sigma) \times I)$ for the triangulated 2-simplex σ of Figure 8.10.

(8.13) Definition. *Let K and L be cell complexes and f and g chain functions from K to L. Then f is chain homotopic to g if there is a chain function $F : K \times I \longrightarrow L$ such that for any simplex $\sigma \in K$*

(1) $F(\sigma \times \{0\}) = f(\sigma)$
(2) $F(\sigma \times \{1\}) = g(\sigma)$.

The function F is a chain homotopy from f to g.

We first show that if $f : |K| \longrightarrow |L|$ is a continuous function and g and h are two simplicial approximations for f, then g and h are chain homotopic. After that, we prove that since they are chain homotopic, it follows that they induce the same homology functions.

(8.14) Lemma. *The simplex $\sigma = \langle v_0, v_1, \ldots v_k \rangle$ is a simplex in K if and only if $\bigcap_{i=0}^{k} st_K(v_i) \neq \emptyset$.*

Proof. If $\sigma = \langle v_0, v_1, \ldots v_k \rangle$ is a simplex in K, then $v_i < \sigma$ for $i = 1, 2, \ldots, k$, so $Int(\sigma) \subseteq st_K(v_i)$ for each i. Therefore, $\bigcap_{i=0}^{k} st_K(v_i) \neq \emptyset$.

If $\bigcap_{i=0}^{k} st_K(v_i) \neq \emptyset$, then let $x \in \bigcap_{i=0}^{k} st_K(v_i) \neq \emptyset$. Since $x \in |K|$, x must lie in the interior of some some cell $\tau \in K$. Therefore, $\tau \cap st_K(v_i) \neq \emptyset$ for each i. By Definition 8.7, $st_K(v_i)$ consists of whole interiors of simplices, so if part of a cell is in $st_K(v_i)$, the whole interior of that cell must lie in $st_K(v_i)$. Thus, $Int(\tau) \subseteq st_K(v_i) = \bigcup_{v_i < \sigma \in K} Int(\sigma)$ for each i. Each of the vertices v_i must then be a vertex of τ. The simplex $\sigma = \langle v_0, v_1, \ldots, v_k \rangle$ is a face of τ and so is a simplex in K. □

(8.15) Lemma. *If $f : |K| \longrightarrow |L|$ is a continuous function, and g and h are two simplicial approximations for f, then for any simplex σ in K there is a simplex $\tau \in L$ such that $g(\sigma)$ and $h(\sigma)$ are both faces of τ.*

▷ **Exercise 8.9.** Let $\sigma = \langle v_0, v_1, \ldots, v_k \rangle$. Prove Lemma 8.15, using Lemma 8.14.

For $g, h : K \longrightarrow L$ and $\sigma = \langle v_0, v_1, \cdots, v_n \rangle$ as in Lemma 8.15, the vertices $g(v_0), h(v_0), g(v_1), h(v_1), \ldots, g(v_k), h(v_k)$ may not be distinct. Some possibilities for the simplex $\tau \in L$ of Lemma 8.15 are pictured in Figures 8.11 and 8.12.

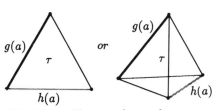

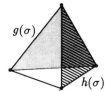

Fig. 8.11. The simplex τ between $g(a)$ and $h(a)$ for an edge a

Fig. 8.12. The simplex τ between $g(\sigma)$ and $h(\sigma)$ for a 2-simplex σ

(8.16) Theorem. *Let K and L be simplicial complexes with $f : |K| \longrightarrow |L|$ a continuous function. If g and h are two simplicial approximations for f, then g and h are chain homotopic.*

Proof. We construct a chain homotopy as in Definition 8.13 between g and h. Let $\sigma = \langle v_0, v_1, \ldots, v_n \rangle$ denote a simplex in K. Recall that $K \times I$ has a simplicial structure with simplices either of the form $\sigma \times \{0\}$ or $\sigma \times \{1\}$ or of the form

$$\sigma \times I = \sum_{k=0}^{n} (-1)^k \langle \widehat{v}_0, \widehat{v}_1, \ldots, \widehat{v}_k, \breve{v}_k, \ldots, \breve{v}_n \rangle.$$

The homotopy $F : K \times I \longrightarrow L$ is defined by defining F on each vertex of the simplicial complex $K \times I$. Define

$$F(v_i \times \{1\}) = F(\widehat{v}_i) = h(v_i)$$
$$F(v_i \times \{0\}) = F(\breve{v}_i) = g(v_i)$$

Therefore,

$$F(\sigma \times \{1\}) = h(\sigma)$$
$$F(\sigma \times \{0\}) = g(\sigma)$$

For a cell of the form $\tau = \langle \widehat{v}_0, \widehat{v}_1, \ldots, \widehat{v}_k, \breve{v}_k, \ldots, \breve{v}_n \rangle$,

$$F(\tau) = \langle h(v_0), h(v_1), \ldots, h(v_k), g(v_k), \ldots, g(v_n) \rangle$$

By Lemma 8.15, $F(\tau)$ is a simplex in L. □

The next theorem proves that if g and h are chain homotopic, then they induce the same homomorphisms on the homology groups. This is important and not too difficult, but very algebraic and abstract, and so is difficult to picture. It is adapted from Cooke and Finney's *Homology of Cell Complexes*.

(8.17) Lemma. *If g and h are chain functions and are chain homotopic, then there are homomorphisms $G_k : C_k(K) \longrightarrow C_{k+1}(L)$ such that $\partial G_{k+1} + G_k \partial = h - g$ for $0 \le k \le dim(K)$.*

Proof. Let g and h be chain homotopic with chain homotopy F, and let σ be a $(k+1)$-cell in K. Define

$$G_{k+1}(\sigma) = F(\sigma \times I).$$

Thus, G_{k+1} takes $(k+1)$-cells in K to $(k+2)$-cells in L, and

$$
\begin{aligned}
\partial G_{k+1}(\sigma) &= \partial F(\sigma \times I) \\
&= F\partial(\sigma \times I) \\
&= F(\sigma \times \{1\} - \sigma \times \{0\} - (\partial(\sigma) \times I)) \\
&= F(\sigma \times \{1\}) - F(\sigma \times \{0\}) - F(\partial(\sigma) \times I) \\
&= h(\sigma) - g(\sigma) - F(\partial(\sigma) \times I) \\
&= h(\sigma) - g(\sigma) - G_k \partial(\sigma)
\end{aligned}
$$

Thus, $\partial G_{k+1} + G_k \partial = h - g$. □

(8.18) Theorem. *If K and L are cell complexes and $g, h : K \longrightarrow L$ are chain homotopic, then $g_* = h_* : H_*(K) \longrightarrow H_*(L)$.*

Proof. Since g and h are chain homotopic, the homomorphisms G_k exist as in the lemma above. Let σ be a k-cycle in K.

$$
\begin{aligned}
h_k(\sigma) - g_k(\sigma) &= h(\sigma) - g(\sigma) \\
&= \partial G_k(\sigma) + G_{k-1}\partial(\sigma)
\end{aligned}
$$

Note that $\partial(\sigma) = 0$, since σ is a cycle. Therefore,

$$h_k(\sigma) - g_k(\sigma) = \partial(G_k(\sigma))$$

However, $\partial(G_k(\sigma))$ is homologous to 0 in $H_k(L)$ since $\partial(G_k(\sigma)) \in B_k(L)$. Thus, $h_k(\sigma) \sim g_k(\sigma)$ for any k-cell σ. It follows that $h_*(C) \sim g_*(C)$ for any chain C, so $h_* = g_* : H_*(K) \longrightarrow H_*(L)$. □

(8.19) Corollary. *If* $f : |K| \longrightarrow |L|$ *is a continuous function, then* $f_* : H_*(K) \longrightarrow H_*(L)$ *is well-defined.*

▷ **Exercise 8.10.**   Prove Corollary 8.19.

▷ **Exercise 8.11.**   Prove that if K, L, and M are simplicial complexes, and $f : |K| \longrightarrow |L|$ and $f' : |L| \longrightarrow |M|$ are continuous functions with g and g' simplicial approximations for f and f' respectively, then $g' \circ g$ is a simplicial approximation for $f' \circ f$.

(8.20) Corollary. *If* K, L, *and* M *are simplicial complexes with continuous functions* $f : |K| \longrightarrow |L|$ *and* $f' : |L| \longrightarrow |M|$, *then* $(f' \circ f)_* = f'_* \circ f_*$.

▷ **Exercise 8.12.**   Prove Corollary 8.20.

(8.21) Corollary. *If If* $f : |K| \longrightarrow |L|$ *is a homeomorphism, then* $f_* : H_*(K) \longrightarrow H_*(L)$ *is an isomorphism.*

▷ **Exercise 8.13.**   Prove Corollary 8.21.

Chapter 9

Homotopy

"The essence of mathematics is not to make simple things complicated, but to make complicated things simple."

Gudder

9.1 Homotopy and homology

It is somewhat discouraging that, after all that work, the euler characteristics and the homology groups cannot tell the difference between the cylinder and the Möbius band. We are not going to try to change that situation, but rather investigate what they have in common that makes them algebraically indistinguishable. We assume, for technical reasons, that all spaces are at least Hausdorff.

(9.1) Definition. *Let X and Y be topological spaces and f and g continuous functions from X to Y. Then f is homotopic to g (written $f \sim g$) if there is a continuous family of continuous functions $f_t : X \longrightarrow Y$ for $0 \le t \le 1$ such that*

    (1) $f_0 = f$
    (2) $f_1 = g$
    (3) $f_t(x)$ *is continuous both as a function of $x \in X$ and as a function of $t \in [0, 1]$.*

The function f_t is a homotopy from f to g.

Another way to represent the homotopy f_t is as the continuous function $F : X \times I \longrightarrow Y$ defined by $F(x, t) = f_t(x)$, where I denotes the unit interval. Think of the t variable in Definition 9.1 as a time variable. One mental picture of a homotopy could be the aging process: Barring appendicitis, aging is a continuous process and the topological shape of an infant is related to the shape of a wrinkled person 90 years old by a homotopy describing the shape at every age between. Another image is a waving flag, topologically equivalent to a rectangle at all times, but the way it is embedded in the 3-dimensional space varies with t.

For an illustration of homotopic functions, let $I = [0, 1]$ be the unit interval, and $f, g : I \longrightarrow Y$ be two functions from the interval to a topological space X as pictured in Figure 9.1. A homotopy from f to g is illustrated in Figure 9.2.

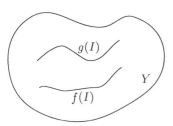
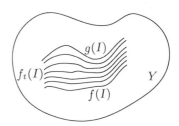

Fig. 9.1. $f, g : I \longrightarrow Y$ **Fig. 9.2.** A homotopy from f to g

As another example, define $f : \mathbb{S}^1 \longrightarrow \mathbb{S}^1$ by

$$f(\cos(2\pi s), \sin(2\pi s)) = \left(\cos\left(2\pi s + \frac{\pi}{2}\right), \sin\left(2\pi s + \frac{\pi}{2}\right)\right)$$

for $0 \le s \le 1$. The function f rotates the circle by $90°$ counterclockwise. A homotopy of f is defined by

$$f_t(\cos(2\pi s), \sin(2\pi s)) = \left(\cos\left(2\pi s + \frac{\pi}{2}t\right), \sin\left(2\pi s + \frac{\pi}{2}t\right)\right)$$

for $0 \le t \le 1$. Note that for each value of t, f_t is a function from the circle to the circle.

$$f_0(\cos(2\pi s), \sin(2\pi s)) = (\cos(2\pi s), \sin(2\pi s))$$
$$f_1(\cos(2\pi s), \sin(2\pi s)) = \left(\cos\left(2\pi s + \frac{\pi}{2}\right), \sin\left(2\pi s + \frac{\pi}{2}\right)\right)$$

Thus, f_0 is the identity function on $\mathbb{S}^1$, denoted by $1_{\mathbb{S}^1}$, and $f_1 = f$. The function f is, therefore, homotopic to the identity function, i.e., $f \sim 1_{\mathbb{S}^1}$.

It may be difficult to appreciate homotopies at first, but they are one of the most useful topological notions. We will use them to explain at last why the cylinder, the Möbius band, and the circle all have the same homology groups, to simplify the computation of homology groups in many cases, and then to invent a whole new algebraic invariant. You will end up liking homotopies. Trust me.

(9.2) Theorem. *If $f_0, f_1 : X \longrightarrow Y$ are homotopic functions, then*

$$f_{0*} = f_{1*} : H_*(X) \longrightarrow H_*(Y)$$

Proof for X and Y triangulable spaces. Let $f_t : X \longrightarrow Y$ be the homotopy from f_0 to f_1, and define $F : X \times I \longrightarrow Y$ by $F(x,t) = f_t(x)$. Then $F(x,1) = f_1(x)$ and $F(x,0) = f_0(x)$ for every $x \in X$, and F is a continuous function. Let K be a simplicial complex on X and L a simplicial structure for Y, so $H_*(X) = H_*(K)$ and $H_*(Y) = H_*(L)$. Let $F' : (K \times I)^{(k)} \longrightarrow L$ be a simplicial approximation for F, where $(K \times I)^{(k)}$ is the kth barycentric subdivision of the canonical simplicial complex on $K \times I$ of Chapter 8.2. Note that taking the barycentric subdivision of $(K \times I)$ gives the barycentric subdivision on the ends, $K \times \{0\}$ and $K \times \{1\}$; i.e., $(K \times I)^{(k)} \cap (K \times \{i\}) = K^{(k)} \times \{i\}$ for $i = 0, 1$. By Theorem 8.2, $H_*(K) \simeq H_*(K^{(k)})$, and by Theorem 8.11 a simplicial approximation F' exists for F with $F(st_{(K \times I)^{(k)}}(v)) \subseteq st_L(F'(v))$ for every vertex in $(K \times I)^{(k)}$. Define $f_0', f_1' : K^{(k)} \longrightarrow L$ by $f_0'(x) = F'(x,0)$ and $f_1'(x) = F'(x,1)$. We first show that f_i' is a simplicial approximation for f_i, $i = 0, 1$. For a vertex $v \in K^{(k)}$,

$$\begin{aligned}
f_i(st_{K^{(k)}}(v)) &= F(st_{(K \times I)^{(k)}}(v \times \{i\}) \cap (K^{(k)} \times \{i\})) \\
&\subseteq F(st_{(K \times I)^{(k)}}(v \times \{i\})) \\
&\subseteq st_L(F'(v \times \{i\})) \\
&= st_L(f_i'(v))
\end{aligned}$$

Therefore, f_i' is a simplicial approximation of f_i and so by Definition 8.12 $f_{i*} = f_{i*}'$. By Definition 8.13, F' is a chain homotopy from f_0' to f_1', and Theorem 8.18 implies that $f_{0*}' = f_{1*}'$. □

Thus, homotopic functions are the same in homology. Next we consider functions which may not be homeomorphisms or topological equivalences, but act like an equivalence up to homotopy.

(9.3) Definition. *If X and Y are topological spaces with $X \subseteq Y$, then a continuous function $r : Y \longrightarrow X$ is retraction if $r(x) = x$ for all $x \in X$.*

In other words, the function r collapses the space Y onto its subset X. Whenever $X \subseteq Y$, define the inclusion function $i : X \longrightarrow Y$ by $i(x) = x$ considered as a point in the bigger space Y. The inclusion function does not move any points, but changes the context in which they are considered. Note that r is a retraction if and only if $r(i(x)) = x$, i.e., $r \circ i = 1_X$. We cannot hope that $i \circ r : Y \longrightarrow Y$ be the identity on Y, since $i \circ r$ first squashes the bigger space Y onto X, and then merely sticks X back into Y without changing it.

(9.4) Definition. *A subset $X \subseteq Y$ is a* deformation retract *of Y if there is a retraction $r : Y \longrightarrow X$ such that $i \circ r \sim 1_Y$. This is denoted by $Y \rightsquigarrow X$. In the case where X is a single point $X = \{x\} \subseteq Y$, Y is called* contractible *and we write $Y \rightsquigarrow *$.*

For example, let $I = [0,1]$ and $X = \{0\} \subseteq [0,1] = Y$. A retraction from I to X is defined by $r(x) = 0$ for each $x \in [0,1] = I$. Note that $i \circ r(x) = i(r(x)) = i(0) = 0$ considered as a point in the interval $[0,1]$. Define $f_t(x) = tx$ for $0 \le t \le 1$ and $x \in I$. Then for each $x \in I$, $f_1(x) = x$ and $f_0(x) = 0 \cdot x = 0 = i \circ r(x)$. Thus, $i \circ r \sim 1_I$ so $\{0\}$ is a deformation retract of I. Picture f_t as shrinking the line segment gradually to a point. Therefore, $I \rightsquigarrow *$.

(9.5) Theorem. *If X is a deformation retract of Y, then $H_*(X) = H_*(Y)$.*

▷ **Exercise 9.1.**   Prove Theorem 9.5.

In Chapter 1 we drew an analogy between homeomorphic spaces in topology and congruent figures in geometry, since congruent figures are geometrically the same and congruent figures have the equal geometrical properties, such as area, length of sides, angle measure, or curvature. The property of being a deformation retract is analogous to similarity in geometry. As with similar figures, if X is a deformation retract of Y, then there are obvious ways that X and Y differ, but there are also some essential characteristics in which they are the same.

Theorem 9.5 often provides an easier way of computing homology groups. For example, consider the space X presented in Figure 9.3.

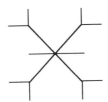

Fig. 9.3. $X \rightsquigarrow *$

The deformation retraction defined above on the line segment can be applied to each leg of X. Thus, $H_k(X) = 0$ for $k > 0$ and $H_0(X) = Z$.

The n-disc $D^n(0,1) = \{\mathbf{x} \in \mathbb{R}^n : \|\mathbf{x}\| < 1\}$ deformation retracts to the origin by a generalization of the retraction of the interval defined above. The retraction on $D^n(0,1)$ is defined by $r(\mathbf{x}) = \mathbf{0}$ with homotopy $f_t(\mathbf{x}) = t\mathbf{x}$. The homotopy is obviously continuous both as a function of t and of $\mathbf{x}$, and $f_0(\mathbf{x}) = 0\mathbf{x} = \mathbf{0} = i(\mathbf{0}) = i \circ r(\mathbf{x})$ and $f_1(\mathbf{x}) = \mathbf{x} = 1_{D^n}$. Thus, $r \circ i = 1_{\{0\}}$

and $i \circ r \sim 1_{D^n}$, so $D^n(0,1) \rightsquigarrow *$. Thus, $H_k(D^n(0,1)) = 0$ for $k > 0$ and $H_0(D^n(0,1)) = Z$. The same reasoning applies to any disc, open or closed, of any radius. Note that the deformation retraction thus defined is identical with the process of radial projection used in Lemma 5.20.

Another deformation retraction takes the square of points (x,y) where $-1 \le x \le 1$, $-1 \le y \le 1$ to the center line $y = 0$, $-1 \le x \le 1$ by defining $r(x,y) = (x,0)$. The homotopy is given by $f_t(x,y) = (x,ty)$ and is illustrated in Figure 9.4.

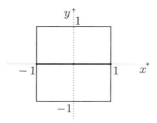

Fig. 9.4. Square $\rightsquigarrow$ line segment

Applying this homotopy to the square of the planar diagram shows that the cylinder deformation retracts to a circle as in Figure 9.5. Thus, $H_k(C) \simeq H_k(\mathbb{S}^1)$.

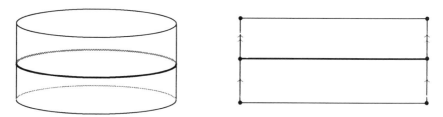

Fig. 9.5. Cylinder $\rightsquigarrow \mathbb{S}^1$

The Möbius band, M, can be retracted onto its meridian circle by the same process (Fig. 9.6), so $H_k(M) \simeq H_k(\mathbb{S}^1)$. This explains why the cylinder and the Möbius band have the same homology groups. The cylinder and the Möbius band are said to have the same *homotopy type* since they both deformation retract to the circle.

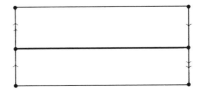

Fig. 9.6. Möbius band $\rightsquigarrow \mathbb{S}^1$

There is a retraction $r : \mathbb{S}^2 \longrightarrow \{(1,0,0)\} \in \mathbb{S}^2$ taking the sphere to a point defined by $r(x,y,z) = (1,0,0)$. This is a retraction since $(1,0,0) \in \mathbb{S}^2$ and $r \circ i(1,0,0) = (1,0,0)$, so $r \circ i = 1_{(1,0,0)}$. If this were a deformation retraction, then $H_*(\mathbb{S}^2) \simeq H_*(\text{point})$, but $H_2(\mathbb{S}^2) = Z \not\simeq H_2(\text{point}) = 0$. Therefore, the sphere does not deformation retract to a point. This retraction does not preserve the essential shape of the sphere. A physical demonstration of this is the fact that if you shrink a balloon down to a point, it will pop.

▷ **Exercise 9.2.** Classify up to homotopy type:

ABCDEFGHIJKLMNOPQRSTUVWXYZ1234567890

▷ **Exercise 9.3.** Compute the homology groups of the "house with two rooms" of Figure 9.7.

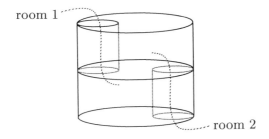

Fig. 9.7. The house with two rooms

▷ **Exercise 9.4.** Compute the homology groups of the solid torus.

▷ **Exercise 9.5.**   Choose a point $x \in \mathbb{T}^2$. Show that the punctured torus $\mathbb{T}^2 - \{x\}$ deformation retracts to $\mathbb{S}^1 \vee \mathbb{S}^1$ and use this fact to compute $H_*(\mathbb{T}^2 - \{x\})$.

▷ **Exercise 9.6.**   Compute the homology groups of each of the standard surfaces with a point removed.

9.2 The fundamental group

Homotopy theory provides another type of algebraic topology besides the homology theory discussed in earlier chapters. Homotopy theory can be used to duplicate many of the results we have already obtained, as well as some new ones.

(9.6) Definition. *A path in X from x to y is a continuous function $\alpha : [0,1] \longrightarrow X$ with $\alpha(0) = x$ and $\alpha(1) = y$.*

Denote the interval $[0,1]$ by I. In this section we will study paths and homotopies of paths. A path is illustrated in Figure 9.8. An arrow along the path indicates the natural direction of α, from $\alpha(0)$ to $\alpha(1)$.

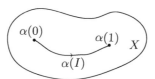

Fig. 9.8. A path α in X

(9.7) Definition. *Let α and β be paths in X from x to y. Then α and β are homotopic paths (denoted $\alpha \sim \beta$) if there is a homotopy $f_t : I \longrightarrow X$ with $f_0 = \alpha$, $f_1 = \beta$ and $f_t(0) = x$ for all t and $f_t(1) = y$ for all t.*

Note that a homotopy of paths from α to β should not move the endpoints. This is an added restriction on the homotopies of the Definition 9.1. A homotopy of paths is illustrated in Figure 9.9, schematically on the left and the image in X on the right.

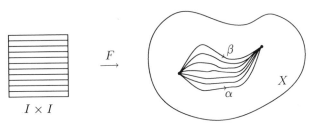

$I \times I$

Fig. 9.9. A homotopy of paths from α to β

(9.8) Theorem. *Homotopy of paths is an equivalence relation on the set of all paths in X from x to y.*

Proof of transitivity. Let α, β, γ be paths in X from x to y, and assume that $\alpha \sim \beta$ and $\beta \sim \gamma$. Then there are homotopies $f_t, g_t : I \longrightarrow X$ such that $f_0 = \alpha$, $f_1 = \beta$, $g_0 = \beta$, $g_1 = \gamma$, and at the endpoints $f_t(0) = x = g_t(0)$ and $f_t(1) = y = g_t(1)$ for all t. A new homotopy F_t can be formed by concatenating these (see Fig. 9.10).

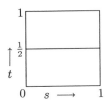

Fig. 9.10. Concatenating homotopies f_t and g_t

This homotopy F_t is defined for $s \in I$ by

$$F_t(s) = \begin{cases} f_{2t}(s), & 0 \leq t \leq \frac{1}{2} \\ g_{2t-1}(s), & \frac{1}{2} \leq t \leq 1 \end{cases}$$

To see that F_t is continuous, note that f_t and g_t are continuous. Using Exercise 3.10, we must show that f_t and g_t agree where they are joined together to form the function F: at $t = \frac{1}{2}$, $F_{\frac{1}{2}}(s) = f_1(s) = \beta(s) = g_0(s)$. Also note that $F_0(s) = f_0(s) = \alpha(s)$ and $F_1(s) = g_1(s) = \gamma(s)$, so F is a homotopy from α to γ. We must show that F is a homotopy of paths, i.e., F is constant at the beginning and end points.

$$F_t(0) = \begin{cases} f_{2t}(0) = x, & 0 \le t \le \frac{1}{2} \\ g_{2t-1}(0) = x, & \frac{1}{2} \le t \le 1 \end{cases}$$

$$F_t(1) = \begin{cases} f_{2t}(1) = y, & 0 \le t \le \frac{1}{2} \\ g_{2t-1}(1) = y, & \frac{1}{2} \le t \le 1 \end{cases}$$

Therefore, $F_t(0) = x$ and $F_t(1) = y$ for all t. Thus, $\alpha \sim \gamma$.   □

▷ **Exercise 9.7.** Prove that homotopy of paths is reflexive and symmetric to complete the proof of Theorem 9.8.

An arithmetic can be defined on the set of paths by joining the paths end to end. This requires that the end of the first path is the beginning of the second.

(9.9) Definition. *If α and β are paths in X with $\alpha(1) = \beta(0)$, define a new path $\alpha * \beta : I \longrightarrow X$ by*

$$\alpha * \beta(s) = \begin{cases} \alpha(2s), & 0 \le s \le \frac{1}{2} \\ \beta(2s - 1), & \frac{1}{2} \le s \le 1 \end{cases}$$

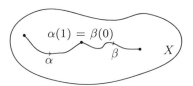

Fig. 9.11. $\alpha * \beta$

The path $\alpha * \beta$, shown in Figure 9.11, is made by running along α at double speed and then running along β at double speed. Since $\alpha(1) = \beta(0)$, the end of the α-path is the beginning of the β-path, so $\alpha * \beta$ is continuous. We use the notation $*$ instead of $+$ since this acts more like multiplication than like addition. In particular, $*$ is not in general commutative, since $\beta * \alpha$ may not even be defined. The following theorem verifies that this multiplication is *well-defined up to homotopy*: i.e., if α and α' are homotopic paths, then $\alpha * \beta$ is homotopic to $\alpha' * \beta$.

(9.10) Theorem. *If $\alpha, \alpha', \beta, \beta'$ are paths in a space X, with $\alpha \sim \alpha'$, $\beta \sim \beta'$, and $\alpha(1) = \beta(0)$, then $\alpha * \beta \sim \alpha' * \beta'$.*

Proof. Since $\alpha \sim \alpha'$, there is a homotopy $f_t : I \longrightarrow X$ with $f_0 = \alpha$, $f_1 = \alpha'$, $f_t(0) = x$, and $f_t(1) = y$. Since $\beta \sim \beta'$, there is a homotopy

$g_t : I \longrightarrow X$ with $g_0 = \beta$, $g_1 = \beta'$, $g_t(0) = x$, $g_t(1) = y$. We must next find a homotopy $F_t : I \longrightarrow X$ satisfying $F_0(s) = \alpha * \beta = f_0 * g_0$ and also $F_1(s) = \alpha' * \beta' = f_1 * g_1$. A natural choice for F_t is the homotopy built from f_t and g_t pictured in Figure 9.12, and defined by

$$F_t(s) = \begin{cases} f_t(2s), & 0 \le s \le \frac{1}{2} \\ g_t(2s - 1), & \frac{1}{2} \le s \le 1 \end{cases}$$

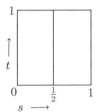

Fig. 9.12. The homotopy F_t

It is easily seen that F_t is the desired homotopy.   □

We next show that this multiplication obeys the usual laws for multiplication:

(9.11) Theorem. *If α, β, and γ are paths in X with $\alpha(1) = \beta(0)$ and $\beta(1) = \gamma(0)$, then $\alpha * (\beta * \gamma) \sim (\alpha * \beta) * \gamma$. Thus, $*$ is associative.*

Proof. The paths $\alpha * (\beta * \gamma)$ and $(\alpha * \beta) * \gamma$ travel over the same ground, but the pacing is different. The path $\alpha * (\beta * \gamma)$ travels along α from 0 to $\frac{1}{2}$, then over $\beta * \gamma$ from $\frac{1}{2}$ to 1, so that β goes from $\frac{1}{2}$ to $\frac{3}{4}$ and γ from $\frac{3}{4}$ to 1. The other path $(\alpha * \beta) * \gamma$ travels over α from 0 to $\frac{1}{4}$, over β from $\frac{1}{4}$ to $\frac{1}{2}$, and then along γ from $\frac{1}{2}$ to 1.

We must find a homotopy $F_t : I \longrightarrow X$ with $F_0(s) = \alpha * (\beta * \gamma)$ and $F_1(s) = (\alpha * \beta) * \gamma$. Construct the straight line homotopy pictured in Figure 9.13. The slanted lines in the square of Figure 9.13 have equations $s = \frac{2-t}{4}$ on the left and $s = \frac{3-t}{4}$ on the right. The leftmost region is $\frac{2-t}{4}$ wide, the center section is $\frac{1}{4}$ wide, and the rightmost section is $1 - \frac{3-t}{4} = \frac{1+t}{4}$ wide. Thus, F_t can be defined by

$$F_t(s) = \begin{cases} \alpha(\frac{4}{2-t}s), & 0 \le s \le \frac{2-t}{4} \\ \beta(4(s - \frac{2-t}{4})), & \frac{2-t}{4} \le s \le \frac{3-t}{4} \\ \gamma((\frac{4}{1+t})(s - \frac{3-t}{4})), & \frac{3-t}{4} \le s \le 1 \end{cases}$$

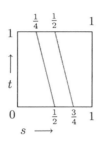

Fig. 9.13. The homotopy F_t

▷ **Exercise 9.8.**   Check that F_t is well-defined, i.e., that both definitions of F_t on the slanted lines of Figure 9.13 match. Show that F_t is a homotopy of paths from $\alpha * (\beta * \gamma)$ to $(\alpha * \beta) * \gamma$.

□

To define an identity for the multiplication, choose a point $x \in X$, and define the *constant path at* x by $e_x(s) = x$ for all $0 \le s \le 1$.

▷ **Exercise 9.9.**   If α is a path in X and $x = \alpha(0)$ and $y = \alpha(1)$, then $e_x * \alpha \sim \alpha$ and $\alpha * e_y \sim \alpha$.

There is not one identity for the multiplication, but rather one defined for each point in the space. Given a path $\alpha : I \longrightarrow X$, define a new path $\overline{\alpha} : I \longrightarrow X$ by $\overline{\alpha}(s) = \alpha(1 - s)$. Note that $\overline{\alpha}(0) = \alpha(1)$ and $\overline{\alpha}(1) = \alpha(0)$. Thus, $\overline{\alpha}$ is the path that travels along α backwards.

▷ **Exercise 9.10.**   If $x = \alpha(0)$ and $y = \alpha(1)$, $\alpha * \overline{\alpha} \sim e_x$ and $\overline{\alpha} * \alpha \sim e_y$.

The path $\overline{\alpha}$ acts as an inverse for α with the multiplication $*$ in homotopy. Thus, we have developed an arithmetic for paths with multiplication which is associative (Theorem 9.11) but not usually commutative. If $\alpha * \beta$ is a product of paths, $\beta * \alpha$ may not even be defined. This multiplication is only defined if the endpoints match up. There is an element (or rather many elements depending on which endpoint is chosen) which acts like multiplication by 1 (Exercise 9.9) and inverses are defined (Exercise 9.10). To eliminate the need to constantly check the endpoints, we can choose a point $x \in X$ and start and end all paths at the point x. This also seems a nice thing to do since we learned in our study of homology that such loops (or cycles) are important to the structure of X.

(9.12) Definition. *A loop in X based at x is a path $\alpha : I \longrightarrow X$ which begins and ends at x; i.e., such that $\alpha(0) = \alpha(1) = x \in X$.*

(9.13) Definition. *Let X be a topological space. The fundamental group or first homotopy group of X is*

$$\pi_1(X, x) = \{all\ loops\ \alpha : I \longrightarrow X\ where\ \alpha\ is\ based\ at\ x\}$$

formed by using the relation $\sim$ given by homotopy of paths instead of $=$.

Elements of $\pi_1(X, x)$ are of the form $[\alpha]$, where α is a loop based at x, and $[\alpha]$ is the set of all loops homotopic to α. The multiplication in $\pi_1(X, x)$ is defined by $[\alpha] \cdot [\beta] = [\alpha * \beta]$.

(9.14) Theorem. *The fundamental group $\pi_1(X, x)$ is a group (in general non-abelian).*

▷ **Exercise 9.11.**   Prove Theorem 9.14.

Since the loops are based at a fixed base point, there is a unique identity element in the fundamental group. The substitution of loops for paths in general removes one difficulty mentioned above, since both $\alpha * \beta$ and $\beta * \alpha$ are always defined, but there is still no reason to assume that they are equal. In general, computations in the fundamental group are somewhat more difficult than the corresponding homology groups, but the invariance of homotopy groups is much easier to prove. Chapter 8 was entirely devoted to proving the invariance of the homology groups. First we show that the homotopy group does not depend on the choice of the base point:

(9.15) Theorem. *Let X be a connected topological space with $x, x' \in X$. Then $\pi_1(X, x) \simeq \pi_1(X, x')$.*

Proof. Choose a path γ from x to x'. This path allows us to define a function $f : \pi_1(X, x) \longrightarrow \pi_1(X, x')$ by

$$f([\alpha]) = [\overline{\gamma} * \alpha * \gamma]$$

This loop is illustrated in Figure 9.14.

The loop $\overline{\gamma} * \alpha * \gamma$ runs backwards along γ from x' to x, then around the loop α, then along γ from x to x', and so is an element of $\pi_1(X, x')$. The function f is well-defined since if $\alpha' \sim \alpha$, then $\overline{\gamma} * \alpha' * \gamma \sim \overline{\gamma} * \alpha * \gamma$ by Theorem 9.10.

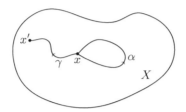

Fig. 9.14. $\overline{\gamma} * \alpha * \gamma$

The function f is a homomorphism since

$$
\begin{aligned}
f([\alpha] \cdot [\beta]) &= f([\alpha * \beta]) \\
&= [\overline{\gamma} * (\alpha * \beta) * \gamma] \\
&= [\overline{\gamma} * \alpha * e_x * \beta * \gamma] \\
&= [\overline{\gamma} * \alpha * \gamma * \overline{\gamma} * \beta * \gamma] \\
&= [\overline{\gamma} * \alpha * \gamma] \cdot [\overline{\gamma} * \beta * \gamma] \\
&= f([\alpha]) \cdot f([\beta]).
\end{aligned}
$$

The inverse of f is easily defined by

$$
f^{-1}([\alpha']) = [\gamma * \alpha' * \overline{\gamma}]
$$

for $[\alpha'] \in \pi_1(X, x')$. Thus, f is an isomorphism.   □

(9.16) Definition. *Let X and Y be topological spaces with a continuous function $f : X \longrightarrow Y$. Choose a base point $x \in X$. If $[\alpha] \in \pi_1(X, x)$ is a loop in X, define $f_*([\alpha]) = [f \circ \alpha] \in \pi_1(Y, f(x))$.*

Note that if $\alpha : I \longrightarrow X$ with $\alpha(0) = x = \alpha(1)$, then $f \circ \alpha : I \longrightarrow Y$ is continuous and $f \circ \alpha(0) = f(x) = f \circ \alpha(1)$. Thus, $f \circ \alpha$ is a loop in Y based at $f(x)$. Any continuous function $f : X \longrightarrow Y$ induces a function $f_* : \pi_1(X, x) \longrightarrow \pi_1(Y, f(x))$ on the fundamental groups.

▷ **Exercise 9.12.** Show that f_* is well-defined; i.e., if $\alpha \sim \alpha'$ in X, then $f \circ \alpha \sim f \circ \alpha'$ in Y.

▷ **Exercise 9.13.** Show that f_* is a homomorphism.

The action of a continuous function, thus, easily translates to an action on the fundamental groups, in stark contrast to the action on homology groups. Let us summarize the relative advantages of the two approaches to algebraic topology:

Setting up the definition: The definition of the homology groups required a lot of machinery (complexes, chains, the homology equivalence relation, etc.), and then was only defined for cellular complexes. This definition can be extended to other spaces, but at an even higher cost; see a more advanced text for the definition of *singular homology*. The homotopy groups are fairly easily defined for any T_2 space.

Group theory involved: The homology groups have the strong advantage of being abelian. Computations in non-abelian groups are more difficult.

Computations: Once the machinery is in place, computations of homology groups are much easier than homotopy groups.

Generalization to higher dimensions: Once homology is defined, it is easily extended to higher dimensions, but the extension of the homotopy groups is less transparent.

Action of a continuous function: The action of a cellular function on the homology groups is not too difficult, though cumbersome, to define, but to define the action of an arbitrary continuous function requires either the Simplicial Approximation Theorem or singular homology. A continuous function easily translates to a homomorphism on the homotopy group.

Efficiency in classifying surfaces: They are equally efficacious in distinguishing the standard surfaces, but working with the non-abelian fundamental group is more difficult than the abelian homology groups. In particular, determining whether two finitely generated groups are isomorphic is much harder in the non-abelian case.

Classification of 3-manifolds: Unsolved.

Since both the fundamental group and the first homology group are defined in terms of loops in the space X, it seems reasonable that they are related. If G is a group, the abelianization of G is the group obtained from the generators and relations of G if one adds commutativity.

(9.17) Theorem. *Let X be a connected topological space. Then $H_1(X)$ is the abelianization of $\pi_1(X)$.*

Thus, the first homology group can tell us nothing that we cannot find from the fundamental group.

That 3-manifolds have proven intractable is shown by the as yet unproven status of one of the oldest and most famous of topological conjectures, made by Henri Poincaré in 1904 [Note that $\pi_1(\mathbb{S}^3) = 0$].

(9.18) Poincaré Conjecture. *If S is a compact, connected 3-manifold without boundary, and $\pi_1(S) = 0$, then S is topologically equivalent to S^3, the 3-sphere.*

Thus, it is unknown whether algebraic topology can detect even the simplest 3-manifold, $\mathbb{S}^3$. We have proven the equivalent homology statement for surfaces (if S is a compact connected surface without boundary and if $H_1(S) = 0$, then S must be topologically equivalent to the sphere $\mathbb{S}^2$). Oddly enough, the analogous statement for higher dimensions has been proved, in 1960 by Stephen Smale for dimensions greater than 6, and soon thereafter extended by John Stallings and Christopher Zeeman for dimensions 5 and 6. In 1981, Michael Freedman proved the 4-dimensional Poincaré conjecture, but the original conjecture remains unsolved.

Chapter 10

Miscellany

10.1 Applications

At this point, it would be a good idea to briefly reread Chapter 2.6, as we now wish to prove higher-dimensional analogs of those theorems. In all of the theorems of that section the key step is the contradiction obtained when a continuous function tries to take a connected space to a non-connected space. This contradiction could be restated in terms of homology groups, since the group $H_0(K)$ detects connectivity. Only with the introduction of functions on topological spaces and their action on the homology groups can we get results comparing two different spaces.

First we recall some results from previous sections. In Examples 6.17 and 6.18, we computed the homology groups of the $(n+1)$-dimensional ball $\mathbb{B}^{n+1}$ and its boundary sphere $\mathbb{S}^n = \partial(\mathbb{B}^{n+1})$. In Definition 9.3, we defined a continuous function $r : Y \longrightarrow X$ to be a retraction if $r(x) = x$ for all $x \in X$. In other words, r collapses the space Y onto the subset X, without moving any of the points of X. The next theorem is a higher-dimensional analog of Theorem 2.30.

(10.1) Theorem. *There does not exist any retraction of the $(n+1)$-ball $\mathbb{B}^{n+1}$ onto its boundary sphere $\mathbb{S}^n = \partial(\mathbb{B}^{n+1})$.*

Proof. Let $i : \mathbb{S}^n \longrightarrow \mathbb{B}^{n+1}$ be the inclusion function, so if $x \in \mathbb{S}^n$, then $i(x) = x \in \mathbb{B}^{n+1}$. The function i does not move any points, but takes the points of $\mathbb{S}^n$ and considers them as points in $\mathbb{B}^{n+1}$. If there were a retraction $r : \mathbb{B}^{n+1} \longrightarrow \mathbb{S}^n$, then $r \circ i(x) = r(x) = x$ for all $x \in \mathbb{S}^n$. Thus, $r \circ i = 1_{\mathbb{S}^n}$.

By the computations in Examples 6.17 and 6.18 of the homology groups of $\mathbb{B}^{n+1}$ and $\mathbb{S}^n$, we have the induced homomorphism

$$i_n : H_n(\mathbb{S}^n) \simeq \mathbb{Z} \longrightarrow H_n(\mathbb{B}^{n+1}) \simeq 0$$

Thus, $i_n(C) = 0$ for all n-cycles $C \in H_n(\mathbb{S}^n)$. The retraction r induces a homomorphism $r_n : H_n(\mathbb{B}^{n+1}) \longrightarrow H_n(\mathbb{S}^n)$. If C is an n-cycle in $\mathbb{S}^n$, then $r_n \circ i_n(C) = r_n(i_n(C)) = r_n(0) = 0$.

However, by Corollary 8.20, $r_n \circ i_n = (r \circ i)_n = (1_{\mathbb{S}^n})_n = 1_{\mathbb{Z}}$. In particular, if C is the non-trivial n-cycle which generates $H_n(\mathbb{S}^n) \simeq \mathbb{Z}$, then $r_n \circ i_n(C) = 1_{\mathbb{Z}}(C) = C$. Thus, there is a contradiction, so a retraction cannot exist. $\qquad\square$

Essentially, this says that one cannot collapse a disc onto its boundary circle without tearing. On the other hand, it was shown in Chapter 9 that one can collapse a disc to a point continuously by shrinking the disc to the center point (which could be thought of as a disc with radius zero). The disc and the circle are not only different shapes, but also too different to be compatible, but the disc and a point are in some sense similar.

▷ **Exercise 10.1.**   Find a retraction of $\mathbb{B}^{n+1} - \mathbf{0}$ to $\mathbb{S}^n$.

(10.2) Brouwer Fixed Point Theorem. *Given any continuous function* $f : \mathbb{B}^n \longrightarrow \mathbb{B}^n$, *there exists a point* $\mathbf{x} \in \mathbb{B}^n$ *with* $f(\mathbf{x}) = \mathbf{x}$; *i.e., any continuous function* f *on the* n-*dimensional ball must have a fixed point.*

Proof. If a continuous function $f : \mathbb{B}^n \longrightarrow \mathbb{B}^n$ exists which does not have a fixed point, then define a new continuous function $g : \mathbb{B}^n \longrightarrow \mathbb{S}^{n-1}$ by considering the ray that starts at the point $f(\mathbf{x}) \in \mathbb{B}^n$ and goes through $\mathbf{x} \in \mathbb{B}^n$. This ray is defined whenever $f(\mathbf{x}) \neq \mathbf{x}$, and intersects the boundary sphere at a unique point as in Figure 10.1.

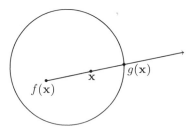

Fig. 10.1. The ray through $\mathbf{x}$ and $f(\mathbf{x})$ defines a new function g

Define $g(\mathbf{x})$ to be the point on $\mathbb{S}^{n-1}$ where this ray crosses the sphere. Note that if $\mathbf{x} \in \mathbb{S}^{n-1}$, then $g(\mathbf{x}) = \mathbf{x}$, so g is a retraction of $\mathbb{B}^n$ onto $\mathbb{S}^{n-1}$. This contradicts Theorem 10.1, so f must have a fixed point. $\qquad\square$

The Brouwer Fixed Point Theorem is an example of what is called an existence proof. It provides no clue as to where or how the fixed point can be found. In his later years, L. E. J. Brouwer became a believer in *constructivism*, a mathematical philosophy which holds that the only valid proofs are constructive ones, and repudiated this theorem.

An amusing illustration of this theorem is:

(10.3) Corollary. *Trace a map or picture. Crumple the copy, being careful not to tear it, and place it so that the crumpled copy lies on the original. There is some point on the copy which lies directly on top of the same point of the original.*

Proof. Note that a rectangular piece of paper is homeomorphic to the ball $\mathbb{B}^2$. Any continuous function on the rectangle must have the fixed point property by Theorem 2.32 and the Brouwer Fixed Point Theorem 10.2. A continuous function from the rectangle to itself is depicted in Figure 10.2, taking a point $\mathbf{x}$ on the original to the point immediately below the image of $\mathbf{x}$ on the crumpled copy.

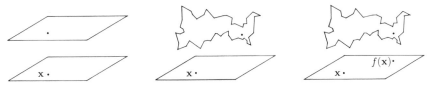

Fig. 10.2. Crumpling a map

This is a composition of two continuous functions: the crumpling and projecting straight downwards.   □

(10.4) Definition. *Let $A \subseteq \mathbb{R}^n$. A vector field $\mathbf{v}$ on A is a continuous function $\mathbf{v}: A \longrightarrow \mathbb{R}^n$. Thus, $\mathbf{v}$ associates to every point $\mathbf{x} \in A$ a vector $\mathbf{v}(\mathbf{x})$ in $\mathbb{R}^n$.*

A vector field $\mathbf{v}$ is illustrated by picturing the set A with the vector $\mathbf{v}(\mathbf{x})$ attached at the point $\mathbf{x} \in A$, as in Figure 10.3. The vector field $\mathbf{v}$ *vanishes* at $\mathbf{x}$ if the vector assigned to $\mathbf{x}$ is zero, i.e., $\mathbf{v}(\mathbf{x}) = 0$.

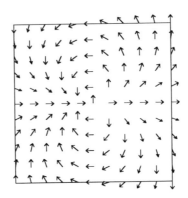

Fig. 10.3. A vector field on the rectangle

(10.5) Corollary. *Any vector field* **v** *on* $\mathbb{B}^2$ *either vanishes at some point or has a point* $\mathbf{x}_0$ *on the boundary* $\mathbb{S}^1$ *where the vector of* **v** *points directly outward and another* $\mathbf{x}_1$ *where* **v** *points directly inward.*

Proof. Assume that $\mathbf{v}(\mathbf{x}) \neq 0$ for all $\mathbf{x} \in B^2$. We will prove that **v** has a point where the vector points directly inward, and another where it points directly outward. Note that any point $\mathbf{x} \in \mathbb{B}^2$ can itself be thought of as a vector. If $\mathbf{x} \in \mathbb{S}^1$, then the vector $\mathbf{v}(\mathbf{x})$ will point directly outward if $\mathbf{v}(\mathbf{x})$ is a multiple of $\mathbf{x}$, i.e., if $\mathbf{v}(\mathbf{x}) = a\mathbf{x}$ for some positive constant a (see Fig. 10.4). Similarly, the vector $\mathbf{v}(\mathbf{x})$ will point directly inward if $\mathbf{v}(\mathbf{x}) = -a\mathbf{x}$ for some positive constant a .

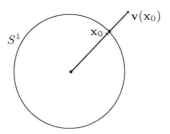

Fig. 10.4. $\mathbf{x}_0 \in \mathbb{S}^1$ with $\mathbf{v}(\mathbf{x}_0)$ pointing directly outward

If **v** never vanishes, we can define a function $f : \mathbb{B}^2 \longrightarrow \mathbb{S}^1 \subseteq \mathbb{B}^2$ by

$$f(\mathbf{x}) = \frac{\mathbf{v}(\mathbf{x})}{\|\mathbf{v}(\mathbf{x})\|}.$$

Note that $f(\mathbf{x})$ is a vector of length 1 and can be thought of as a point on $\mathbb{S}^1$. The function f is continuous if $\mathbf{v}(\mathbf{x}) \neq 0$ for all $\mathbf{x}$. The Brouwer Fixed Point Theorem 10.2 implies that f has a fixed point, so there is an $\mathbf{x}_0$ such that $f(\mathbf{x}_0) = \mathbf{x}_0$. The point $\mathbf{x}_0$ must, thus, be on the circle $\mathbb{S}^1$, and $\mathbf{v}(\mathbf{x}_0) = \|\mathbf{v}(\mathbf{x}_0)\| \cdot \mathbf{x}_0$, so $\mathbf{v}(\mathbf{x}_0)$ points outward. The point $\mathbf{x}_1$ where the vector field points directly inward is found by applying the same argument to the function

$$g(\mathbf{x}) = -\frac{\mathbf{v}(\mathbf{x})}{\|\mathbf{v}(\mathbf{x})\|}$$

If $\mathbf{v}$ does not vanish, there are points such as $\mathbf{x}_0$ and $\mathbf{x}_1$. Therefore, either $\mathbf{v}$ vanishes at some point or there are points on the boundary $\mathbb{S}^1$ where the vector points directly outward and inward. □

(10.6) Corollary. *Everyone has at least one whirl or cowlick on his head.*

▷ **Exercise 10.2.** Prove Corollary 10.6.

Further results about vector fields will be developed in the next chapter. We next turn our attention to functions on spheres.

(10.7) Definition. *A continuous function $f : \mathbb{S}^n \longrightarrow \mathbb{S}^m$ is called antipode-preserving if $f(-\mathbf{x}) = -f(\mathbf{x})$ for each $x \in \mathbb{S}^n$.*

A function f is antipode-preserving if opposite points remain opposite under the action of f. An example of an antipode-preserving function is $f : S^2 \longrightarrow S^2$ which rotates the sphere about the Z-axis by $90°$, as in Figure 10.5.

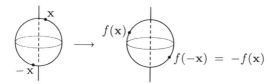

Fig. 10.5. An antipode-preserving function

(10.8) Borsuk-Ulam Theorem. *There is no continuous antipode-preserving continuous function $f : \mathbb{S}^2 \longrightarrow \mathbb{S}^1$.*

Proof. Assume that, contrary to the hypothesis, there is such a function $f : \mathbb{S}^2 \longrightarrow \mathbb{S}^1$ with $f(-\mathbf{x}) = -f(\mathbf{x})$, and that f is cellular. In Example 7.17, we defined the double covering function $p : \mathbb{S}^2 \longrightarrow \mathbb{P}^2$ which glued together antipodal points on the sphere $\mathbb{S}^2$ to get the projective plane $\mathbb{P}^2$. Give $\mathbb{S}^2$ and

$\mathbb{P}^2$ the complex structures of Figure 10.6. and note that $p(\sigma) = p(\tau) = \rho$, $p(a) = p(b) = c$, and $p(P) = p(Q) = R$, so p is a cellular function.

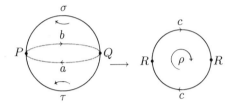

Fig. 10.6. Complex structures for $\mathbb{S}^2$ and $\mathbb{P}^2$

A point in $\mathbb{P}^2$ can be considered as the image of a pair of antipodal points in $\mathbb{S}^2$, since $p(-\mathbf{x}) = p(\mathbf{x})$ for each $\mathbf{x} \in \mathbb{S}^2$.

Another cellular function $q : \mathbb{S}^1 \longrightarrow \mathbb{S}^1$ is defined by identifying antipodal points on the circle $\mathbb{S}^1$, as pictured in Figure 10.7.

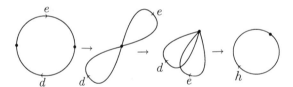

Fig. 10.7. Identifying antipodal points in $\mathbb{S}^1$

Note that $q(\mathbb{S}^1) = \mathbb{S}^1$ and $q(x) = q(-x)$ for any point $x \in \mathbb{S}^1$. Also, for the 1-chain $d + e$, $q(d + e) = 2h$, where h is the generating loop of $\mathbb{S}^1$. Define yet another function $g : \mathbb{P}^2 \longrightarrow \mathbb{S}^1$ by $g(\mathbf{x}) = qf(\mathbf{x})$ for $\mathbf{x} \in \mathbb{S}^2$. Since $qf(-\mathbf{x}) = q(-f(\mathbf{x})) = qf(\mathbf{x})$, the function g is defined on $\mathbb{P}^2$, which is $\mathbb{S}^2$ with opposite points identified.

Thus, the diagram below commutes:

$$
\begin{array}{ccc}
\mathbb{S}^2 & \xrightarrow{\ f\ } & \mathbb{S}^1 \\
{\scriptstyle p}\downarrow & & \downarrow{\scriptstyle q} \\
\mathbb{P}^2 & \xrightarrow[\ g\ ]{} & \mathbb{S}^1
\end{array}
$$

Theorem 7.5 implies that the commutative diagram on the topological spaces induces another commutative diagram on the groups of 1-chains:

$$C_1(\mathbb{S}^2) = [[a, b]] \xrightarrow{f_1} C_1(\mathbb{S}^1) = [[d, e]]$$

$$p_1 \downarrow \qquad\qquad\qquad \downarrow q_1$$

$$C_1(\mathbb{P}^2) = [[c]] \xrightarrow{\quad g_1 \quad} C_1(\mathbb{S}^1) = [[f]]$$

Consider the 1-chain a in $\mathbb{S}^2$, pictured in Figure 10.6. Note that a is not a 1-cycle in $\mathbb{S}^2$ since $\partial(a) = Q - P$. However $c = p(a)$ is a 1-cycle in $\mathbb{P}^2$ since $\partial(c) = R - R = p(Q) - p(P)$. Thus, $p(a)$ can be viewed as an element in $H_1(\mathbb{P}^2) = \mathbb{Z}/2$. The function g induces a homomorphism $g_1 : H_1(\mathbb{P}^2) \simeq \mathbb{Z}/2 \longrightarrow H_1(\mathbb{S}^1) \simeq \mathbb{Z}$. As noted in the Appendix, the only homomorphism from $\mathbb{Z}/2$ to $\mathbb{Z}$ is the zero homomorphism, so $g_1 = 0$. Thus, $g(p(a)) \sim \emptyset$ in $\mathbb{S}^1$, and so $g(p(a))$ is a trivial loop in homology. The only trivial loops on the circle are those which fail to complete a circuit around the circle, since if a loop goes all the way around the circle, then the loop will be homologous to some nonzero multiple of the generator h. Thus, the loop $g(p(a))$ either starts a circuit and then doubles back on itself, or never goes anywhere at all.

On the other hand, $g(p(a)) = q(f(a))$. The endpoints of a are the antipodal points P and Q. Since the function f preserves antipodal points, $f(P)$ and $f(Q)$ are antipodal in $\mathbb{S}^1$. Therefore, $f(a)$ cannot be a loop or 1-cycle in $\mathbb{S}^1$. The 1-chain $f(a)$ must either stretch halfway around the circle or loop around the circle a while and then go halfway around the circle, since the endpoints are antipodal. However, any 1-chain between a pair of antipodal points will form a circuit of $\mathbb{S}^1$ after antipodal points are glued together by the function q. Thus, $q(f(a))$ is a non-trivial 1-cycle in $\mathbb{S}^1$. This contradicts our argument that $g(p(a))$ is a trivial 1-cycle on $\mathbb{S}^1$. Therefore, functions such as f cannot exist. □

This theorem, like Theorem 10.1, says that the sphere and the circle are too different to be able to support an antipode-preserving continuous function. There are a number of corollaries to this seemingly dull theorem.

(10.9) Corollary. If $f : \mathbb{S}^2 \longrightarrow \mathbb{R}^2$ is a continuous function such that $f(-\mathbf{x}) = -f(\mathbf{x})$ for all x, then there is a point $\mathbf{y} \in \mathbb{S}^2$ with $f(\mathbf{y}) = (0, 0)$.

Proof. If a point such as $\mathbf{y}$ did not exist, we could define a continuous function by

$$g(\mathbf{x}) = \frac{1}{\|f(\mathbf{x})\|} f(\mathbf{x})$$

Since $\|g(\mathbf{x})\| = 1$, g is a function from $\mathbb{S}^2$ to $\mathbb{S}^1$. Note that

$$g(-\mathbf{x}) = \frac{1}{\|f(-\mathbf{x})\|} f(-\mathbf{x}) = \frac{1}{\|-f(\mathbf{x})\|}(-f(\mathbf{x})) = -\frac{1}{\|f(\mathbf{x})\|} f(\mathbf{x}) = -g(\mathbf{x})$$

so g is antipode-preserving, contradicting Theorem 10.8. Thus, f must have a point $\mathbf{y} \in \mathbb{S}^2$ with $f(\mathbf{y}) = (0,0)$. $\square$

(10.10) Corollary. *If $f : \mathbb{S}^2 \longrightarrow \mathbb{R}^2$ is a continuous function, then there is a point $\mathbf{y} \in \mathbb{S}^2$ with $f(\mathbf{y}) = f(-\mathbf{y})$.*

Proof. Given any continuous function $f : \mathbb{S}^2 \longrightarrow \mathbb{R}^2$, define a continuous function $g : \mathbb{S}^2 \longrightarrow \mathbb{R}^2$ by $g(\mathbf{x}) = f(\mathbf{x}) - f(-\mathbf{x})$. It then follows that $g(-\mathbf{x}) = f(-\mathbf{x}) - f(\mathbf{x}) = -g(\mathbf{x})$. By Corollary 10.9, there is a point $\mathbf{y}$ with $g(\mathbf{y}) = (0,0)$, and so $f(\mathbf{y}) = f(-\mathbf{y})$. $\square$

(10.11) Corollary. *If $\mathbb{S}^2$ is divided into three closed subsets, then at least one of these subsets contains a pair of antipodal points.*

Proof. Let $\mathbb{S}^2$ be divided into three closed sets, so $\mathbb{S}^2 = A_1 \cup A_2 \cup A_3$. Define continuous functions $f_i : \mathbb{S}^2 \longrightarrow \mathbb{R}$ for $i = 1$ and 2, by $f_i(\mathbf{x}) = \mathrm{dist}(\mathbf{x}, A_i)$ where $\mathrm{dist}(\mathbf{x}, A_i)$ measures the distance between the point $\mathbf{x}$ and the nearest point of the closed set A_i. Note that since f_i is a distance function, for any point $\mathbf{y} \in \mathbb{S}^2$, $f_i(\mathbf{y}) = f_i(-\mathbf{y}) \geq 0$. A function $f : \mathbb{S}^2 \longrightarrow \mathbb{R}^2$ is defined by $f(\mathbf{x}) = (f_1(\mathbf{x}), f_2(\mathbf{x}))$. By Corollary 10.10, there must be a point $\mathbf{y} \in \mathbb{S}^2$ where $f(\mathbf{y}) = f(-\mathbf{y})$. Therefore, $f_i(\mathbf{y}) = f_i(-\mathbf{y})$ for $i = 1$ and 2. If either $f_i(\mathbf{y}) = f_i(-\mathbf{y}) = 0$, then both $\mathbf{y}$ and $-\mathbf{y}$ are in A_i. If $f_i(\mathbf{y}) = f_i(-\mathbf{y}) > 0$ for $i = 1$ and 2, then neither $\mathbf{y}$ nor $-\mathbf{y}$ are in the sets A_1 and A_2, so both must lie in A_3. $\square$

▷ **Exercise 10.3.**   Prove that no subset of $\mathbb{R}^2$ is homeomorphic to $\mathbb{S}^2$.

▷ **Exercise 10.4.**   Prove that there does not exist a continuous 1-1 function $f : \mathbb{R}^3 \longrightarrow \mathbb{R}^2$.

(10.12) Corollary. *At any time, there are two points directly opposite each other on the earth with exactly the same temperature and barometric pressure (or any other two continuous functions).*

▷ **Exercise 10.5.**   Prove Corollary 10.12.

The following theorem is the 3-dimensional analog of the Pancake Theorem 2.36. The sets A, B, and C need not be connected. Consider A as the bread, B the ham, and C the cheese.

(10.13) Ham Sandwich Theorem. *Let A, B, and C be compact subsets of $\mathbb{R}^3$. There is a plane which divides each region exactly in half (by volume).*

Proof. Since A, B, and C are bounded, we may assume that they lie within the sphere of diameter 1, since we can change the scale if necessary. For each point $\mathbf{x} \in \mathbb{S}^2$, let $D_{\mathbf{x}}$ be the diameter line through $\mathbf{x}$. For t with $0 \leq t \leq 1$, let P_t be the plane perpendicular to $D_{\mathbf{x}}$ at distance t from the point $\mathbf{x}$ (see Fig. 10.8).

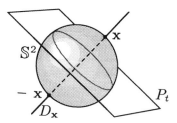

Fig. 10.8. The diameter $D_{\mathbf{x}}$ and the plane P_t

Note that P_t divides A into two (probably not equal) parts. Let

$$f_{\mathbf{x},1}(t) = \text{volume of the portion of } A \text{ on the } \mathbf{x}\text{-side of } P_t$$
$$f_{\mathbf{x},2}(t) = \text{volume of the portion of } A \text{ on the far side of } P_t$$

Note that $f_{\mathbf{x},1}(0) = 0$ and $f_{\mathbf{x},1}(1) = \text{vol}(A)$. Similarly, $f_{\mathbf{x},2}(0) = \text{vol}(A)$ and $f_{\mathbf{x},2}(1) = 0$. Define $f_{\mathbf{x}}(t) = f_{\mathbf{x},1}(t) - f_{\mathbf{x},2}(t)$.

The function $f_{\mathbf{x}} : [0,1] \longrightarrow \mathbb{R}$ is continuous. We wish to find a point $s_{\mathbf{x}}$, $0 \le s_{\mathbf{x}} \le 1$, where $f_{\mathbf{x}}(s_{\mathbf{x}}) = 0$. The plane $P_{s_{\mathbf{x}}}$ will then divide A into two equal parts. If $f_{\mathbf{x}}(t)$ were never zero, then one could define another continuous function $h : [0,1] \longrightarrow \{-1, +1\}$ by

$$h(t) = \frac{f_{\mathbf{x}}(t)}{|f_{\mathbf{x}}(t)|}$$

As in Theorem 2.36, this yields a continuous function taking a connected set onto a non-connected set, which is a contradiction unless h is not continuous, i.e., unless there is a point $s_{\mathbf{x}}$ where $f_{\mathbf{x}}(s_{\mathbf{x}}) = 0$. Define a continuous function $\alpha : \mathbb{S}^2 \longrightarrow \mathbb{R}$ by $\alpha(\mathbf{x}) = s_{\mathbf{x}}$, the point on $D_{\mathbf{x}}$ where the plane $P_{\alpha(\mathbf{x})}$ bisects A. Note that for any $\mathbf{x} \in \mathbb{S}^2$, $\alpha(-\mathbf{x}) = 1 - \alpha(\mathbf{x})$. Similarly, find a point $\beta(\mathbf{x})$ where $P_{\beta(\mathbf{x})}$ bisects B and and $\gamma(\mathbf{x})$ where $P_{\gamma(\mathbf{x})}$ bisects C. Define a function $g : \mathbb{S}^2 \longrightarrow \mathbb{R}^2$ by

$$g(\mathbf{x}) = (\alpha(\mathbf{x}) - \beta(\mathbf{x}), \alpha(\mathbf{x}) - \gamma(\mathbf{x}))$$

Note that

$$\begin{aligned}
g(-\mathbf{x}) &= (\alpha(-\mathbf{x}) - \beta(-\mathbf{x}), \alpha(-\mathbf{x}) - \gamma(-\mathbf{x})) \\
&= ((1 - \alpha(\mathbf{x})) - (1 - \beta(\mathbf{x})), (1 - \alpha(\mathbf{x})) - (1 - \gamma(x))) \\
&= (\beta(\mathbf{x}) - \alpha(\mathbf{x}), \gamma(\mathbf{x}) - \alpha(\mathbf{x}))
\end{aligned}$$

$$= -g(\mathbf{x})$$

By Corollary 10.9, there is a point $\mathbf{y} \in \mathbb{S}^2$ with $g(\mathbf{y}) = (0,0)$. Thus, at the point $\mathbf{y}$, $\alpha(\mathbf{y}) = \beta(\mathbf{y})$ and $\alpha(\mathbf{y}) = \gamma(\mathbf{y})$. Therefore, $P_{\alpha(\mathbf{y})} = P_{\beta(\mathbf{y})} = P_{\gamma(\mathbf{y})}$ and this plane bisects all three regions.   □

10.2 The Jordan Curve Theorem

The *Jordan Curve Theorem*, first misproved by Camille Jordan in 1887, is typical of topological facts which initially seem obvious but require surprisingly long and difficult proofs. We will prove the theorem for polygonal curves only. A polygonal curve is a 1-complex made up of straight edges and vertices. The following proof is adapted from Moise's *Geometric Topology in Dimensions 2 and 3*.

(10.14) Jordan Curve Theorem. *Let $C \subseteq \mathbb{R}^2$ be topologically equivalent to the circle. Then $\mathbb{R}^2 - C$ has two components: one bounded and the other unbounded.*

Proof for C a polygonal 1-complex. Let $C \subseteq \mathbb{R}^2$ be a 1-complex topologically equivalent to $\mathbb{S}^1$. Then C consists of a finite number of edges a_i and vertices v_i, $i = 1, 2, \ldots, n$, and each edge is a line segment. Let P be an arbitrary point in $\mathbb{R}^2 - C$. Choose a line ℓ such that ℓ is not parallel to any of the edges a_i, and $v_i \notin \ell$ for each i, as in Figure 10.9.

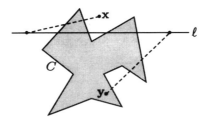

Fig. 10.9. The polygonal 1-complex C, the line ℓ, an inside point $\mathbf{y}$, and an outside point $\mathbf{x}$

    The intersection of the 1-chain C and the line ℓ is a finite number of points, so $\ell - C$ will be a finite union of disjoint intervals, some of which must be unbounded since ℓ is unbounded. A point $\mathbf{x} \in \mathbb{R}^2 - C$ is defined to be *outside* of C if either one of the following conditions hold:

(1) $\mathbf{x}$ is in an unbounded component of $\ell - C$
(2) to get from $\mathbf{x}$ to an unbounded component of $\ell - C$, one must cross C an even number of times. Note that 0 is considered to be even.

Define $\mathbf{y}$ to be *inside* C if to get from $\mathbf{y}$ to an unbounded component of $\ell - C$, one must cross C an odd number of times. Let A be the set of all points of $\mathbb{R}^2 - C$ which are outside of C, and let B be the set of all points inside of C. Every point of $\mathbb{R}^2 - C$ will lie either in A or in B, so $\mathbb{R}^2 - C = A \cup B$ and $A \cap B = \emptyset$. It seems intuitively clear that if it takes n crossings to get from a point $\mathbf{x}$ to an unbounded component of $\ell - C$, then it will also take n crossings to go from a point $\mathbf{z}$ very very close to $\mathbf{x}$ to an unbounded component of $\ell - C$. Thus, if $\mathbf{x} \in A$, then $\mathbf{z} \in A$ for all $\mathbf{z}$ close to $\mathbf{x}$. Therefore, the set A is open. A similar argument shows that B is open. Thus, $\mathbb{R}^2 - C$ is not connected by Exercise 2.29 and has at least two components. Furthermore, A is unbounded since it contains an unbounded component of $\ell - C$.

It remains to show that A and B are each connected, and so $\mathbb{R}^2 - C$ has exactly two components. One way of showing that A is connected is to show that any two points of A can be connected by a path lying entirely in A.

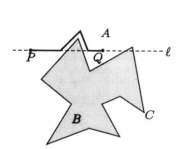

Fig. 10.10. Canceling a pair of crossings

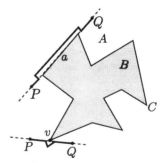

Fig. 10.11. Detouring around a vertex or edge

Let P and Q be two points in A, and consider the line $\ell = PQ$. If ℓ is neither collinear with any of the edges a_i of C nor passes through any of the vertices v_i, then since it takes an even number of crossings to get from P to an unbounded section of $\ell - C$, and an even number of crossings to get from Q to the same unbounded interval, there are an even number of crossings to get from P to Q. These crossings can be eliminated in pairs, by replacing the section of the line segment PQ that passes through the set B of inside points by a parallel translate of that section of C lying entirely in A as illustrated in Figure 10.10.

If $\ell = PQ$ happens to pass through vertex v_i or to include edge a_i, one can detour inside of A around the vertex or edge, as in Figure 10.11. Similarly, one can connect any two points lying in B, so A and B are connected and, thus, $\mathbb{R}^2 - C$ has two connected components.   □

Notice that the proof above is not quite satisfactory due to a blatant appeal to intuition in the second paragraph, and also a failure to prove that the definition of inside and outside does not depend on the the choice of the line ℓ but is enough to give an idea of the procedure. Lest we become too complacent, consider the snowflake curve defined as the limit of the curves S_n. Let S_1 be an equilateral triangle with side 1, and remove the middle thirds of each side and build equilateral triangles in their places to create S_2. Continue removing the middle thirds of each side and replacing them by triangles to get S_n. The first four steps are pictured in Figure 10.12. The snowflake curve S is the limit of the S_n's.

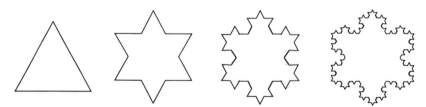

Fig. 10.12. The first four steps of the snowflake curve

The Jordan Curve Theorem is true for S, but it is much harder to see which points (especially those near the boundary) are inside and which outside.

The Jordan Curve Theorem is true for any curve C homeomorphic to $\mathbb{S}^1$ in $\mathbb{R}^2$, and also generalizes to higher dimensions, but this we do not prove.

(10.15) Generalized Jordan Curve Theorem. *If $S \subseteq \mathbb{R}^n$ and S is topologically equivalent to the sphere $\mathbb{S}^{n-1}$, then $\mathbb{R}^n - S$ has two components: one bounded, the other unbounded.*

Extensions of these theorems, which also seem intuitively obvious but require even more subtle proof (which we omit), follow. The first of these was proven by Schoenflies (with corrections by Brouwer) in 1906, and the generalization by Morton Brown in 1960.

(10.16) Schoenflies Theorem. *Let $C \subseteq \mathbb{R}^2$ be a Jordan curve with a homeomorphism $f : \mathbb{S}^1 \longrightarrow C \subseteq \mathbb{R}^2$. Then f can be extended to a homeomorphism $\bar{f} : \mathbb{R}^2 \longrightarrow \mathbb{R}^2$.*

If A is the unbounded component of $\mathbb{R}^2 - C$ and B the bounded component, then the Schoenflies Theorem implies that $B \cup C$ is homeomorphic to the disc D^2, and $A \cup C$ is homeomorphic to $\mathbb{R}^2 - Int(D^2)$.

(10.17) Generalized Schoenflies Theorem. *Let $S \subseteq \mathbb{R}^n$ be a finite complex with a homeomorphism $f : \mathbb{S}^{n-1} \longrightarrow S \subseteq \mathbb{R}^n$. Then f can be extended to a homeomorphism $\bar{f} : \mathbb{R}^n \longrightarrow \mathbb{R}^n$.*

The Schoenflies Theorem is not true for arbitrary topological spaces, since examples such as the Alexander horned sphere exist; see Figure 10.13.

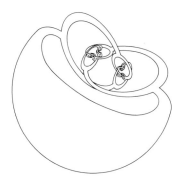

Fig. 10.13. The third stage of the Alexander horned sphere

Start with $\mathbb{S}^2$ and push out two horns or arms. This is a homeomorphism. Now keep all but the ends of the arms fixed, and push out two fingers or claws and almost link them. This again is a homeomorphism. Let the fingers grow little fingers (fingerlets?) and almost link them. The limit of this sequence is a homeomorphism since the action takes place inside a very small neighborhood, and the sequence of continuous functions is *uniformly convergent*. Uniform convergence is studied in advanced calculus or real analysis courses, where it is shown that the uniform limit of continuous functions is continuous. Thus, the Alexander horned sphere is homeomorphic to $\mathbb{S}^2$.

If the Schoenflies Theorem were true, then the region exterior to the Alexander horned sphere, which we will denote by $\mathbb{R}^3 - \mathbb{A}$, would be homeomorphic to $\mathbb{R}^3 - \mathbb{B}^3$. Note that any loop in $\mathbb{R}^3 - \mathbb{B}^3$ can be deformation retracted to a point, with the retraction staying inside $\mathbb{R}^3 - \mathbb{B}^3$, as in Figure 10.14. Thus, $\pi_1(\mathbb{R}^3 - \mathbb{B}^3) = 0$.

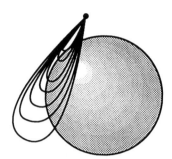

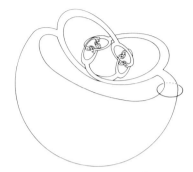

Fig. 10.14. Any loop deforms to a point in $\mathbb{R}^3 - \mathbb{B}^3$

Fig. 10.15. This loop in $\mathbb{R}^3 - \mathbb{A}$ cannot be deformed to a point

On the other hand, the loop encircling one of the arms of the Alexander horned sphere (Fig. 10.15) cannot be deformed to a point in $\mathbb{R}^3 - \mathbb{A}$, since it cannot be extricated from the web of horns. This dilemma means that $\pi_1(\mathbb{R}^3 - \mathbb{A}) \neq 0$, and so $\mathbb{R}^3 - \mathbb{B}^3$ is not homeomorphic to $\mathbb{R}^3 - \mathbb{A}$. However, the theorem is true for finite complexes, which is the case we will use in the next section.

10.3 3-Manifolds

A 3-manifold is locally like $\mathbb{R}^3$. The solid ball is an example of a 3-manifold with boundary. The 3-sphere, $\mathbb{S}^3$, can be constructed from the 3-ball, $\mathbb{B}^3$, by crushing its boundary to a point as in Example 6.18. Another 3-manifold is the 3-torus, $\mathbb{T}^3$, which will be constructed in Example 10.18 by taking a cube and identifying the six sides in pairs, so that opposite faces are glued together.

▷ **Exercise 10.6.** Stand in a cubical room and imagine what would happen if you pushed the right-hand wall out and around and glued it to the left-hand wall, then glued the front and back walls together, and finally glued the ceiling and the floor together. Looking straight ahead, what would you see? What would you see looking up? What happens if you go for a walk?

A very natural example of a 3-manifold is the universe. I grew up in a family that was not particularly intellectual — not illiterate, but we did not discuss global issues around the supper table. I still remember the distinct shock of surprise that I felt when my first grade teacher mentioned, in an unwarrantably off-hand manner, that of course the earth was round. We have no more reason to think that the universe is $\mathbb{R}^3$ than our ancestors (and my preschool self) had to think the earth flat. As far as we know, the universe is homogeneous, or locally 3-dimensional. Thus, as best as we can tell, the universe is a 3-manifold. It could be $\mathbb{R}^3$, or a 3-sphere or a 3-torus, or any other 3-manifold, some compact and others not, some orientable and some nonorientable. The classification of 3-manifolds is the most famous unsolved problem in topology.

Any 3-manifold can be built from a set of solid polyhedra with faces identified in pairs, just as surfaces were built from triangles. Consider the cube

$$C = \{(x, y, z) : -1 \le x \le 1, \ -1 \le y \le 1, \ -1 \le z \le 1\}$$

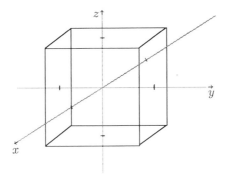

Fig. 10.16. The cube C

where x runs back to front, y left to right, and z down to up. Thus, the points of the form $(1, y, z)$ lie on the front face of the cube, and $(-1, y, z)$ on the back face. Similarly, $(x, -1, z)$ lies on the left face of C, etc. See Figure 10.16.

(10.18) Example. Glue points of the form $(-1, y, z)$ to the corresponding point $(1, y, z)$, $(x, -1, z)$ to $(x, 1, z)$, and $(x, y, -1)$ to $(x, y, 1)$. The result is a 3-complex, and we ask if it is a 3-manifold. We must verify that every point has a neighborhood topologically equivalent to the open ball. Points inside the cube clearly have neighborhoods homeomorphic to a ball. Points on the faces have half-ball neighborhoods which are glued together to form a solid ball, as in Figure 10.17.

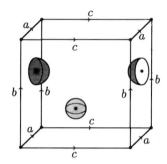

Fig. 10.17. Neighborhoods of an interior point and a face point

A point on an edge, for example **x** in Figure 10.18, has a neighborhood consisting of four 1/4-balls.

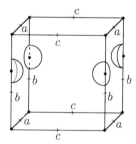

Fig. 10.18. Neighborhood of an edge point

To see that this neighborhood is the 3-ball, we use the Generalized Schoenflies Theorem 10.17 and show that its boundary is homeomorphic to the sphere. The boundary of this neighborhood consists four pieces that each look like the skin of a section of an orange which are glued together as in Figure 10.19. The appendant edges are indicated to help decide how to identify the edges of the boundary complex. Note that section (1) is glued to (2) along d on the ab face and to (4) along e on the bc face, etc. Thus, the boundary of the neighborhood can be glued together to form an sphere, so the neighborhood itself is topologically equivalent to a ball.

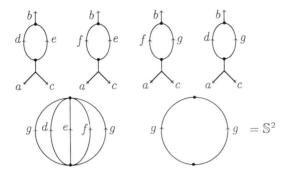

Fig. 10.19. The boundary of the neighborhood of the edge point is the sphere

Consider next the vertices of the cube (Fig. 10.20).

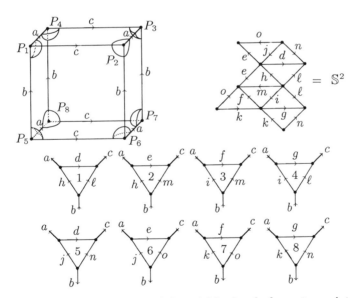

Fig. 10.20. The boundary of the neighborhood of a vertex point

Note that the gluing instructions indicate that all eight corners of the box are to be identified, since $(-1, 1, 1) \sim (1, 1, 1) \sim (1, -1, 1) \sim (1, -1, -1) \sim$

$(1, 1, -1) \sim (-1, 1, -1) \sim (-1, -1, -1) \sim (-1, -1, 1)$. Thus, a neighborhood of the vertex P consists of eight 1/8-balls glued together as seen in Figure 10.20. The boundary of this neighborhood is formed by gluing eight triangles together. The boundary of the neighborhood is a surface (since it is formed by identifying edges of a set of triangles in pairs), which can be seen by the euler characteristic, by cutting and pasting, or by homology groups, to be $\mathbb{S}^2$. By Theorem 10.17, the neighborhood is a ball. Thus, the space above has ball neighborhoods for each point and so is a 3-manifold. It is commonly called the 3-torus or $\mathbb{T}^3$, since the construction is analogous to the standard 2-dimensional torus, $\mathbb{T}^2$.

(10.19) Example. Another example is created by taking the cube and identifying points of the form $(1, y, z)$ to $(-1, -y, z)$, $(x, 1, z)$ to $(x, -1, -z)$, and $(x, y, 1)$ to $(-x, y, -1)$. Since $(x, 1, z)$ is glued to $(x, -1, -z)$, this complex glues the left wall (where $y = -1$) to the right ($y = 1$) but with a flip.

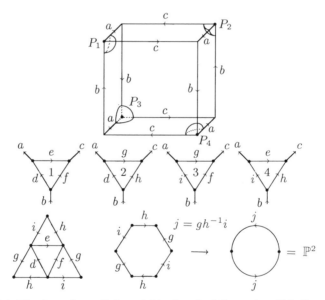

Fig. 10.21. The boundary of the neighborhood of the vertex P is the projective plane

Imagine pushing the walls together with your hands. In the previous example of $\mathbb{T}^3$ you would bring your hands together palm to palm, with the fingers of both hands pointing up. In this example, you would push the walls out and bring your hands together so that the palm of one hand meets the back of the other, with the fingers of one hand pointing up and the other down. The other gluings are similar.

Point P_1 in Figure 10.21 has coordinates $(1, -1, 1)$. Thus, P_1 gets glued to $P_2 = (-1, 1, 1)$ which gets glued to $P_3 = (-1, -1, -1)$ which gets glued to $P_4 = (1, 1, -1)$. A neighborhood of the point P formed by gluing together P_1, P_2, P_3, P_4 is built from four 1/4-balls. The boundary of the neighborhood at P is the cell complex of Figure 10.21. The boundary of the neighborhood of the vertex P is $\mathbb{P}^2$, not $\mathbb{S}^2$, so the neighborhood is not topologically equivalent to a ball. The figure is then not a manifold, but what is called a *pseudomanifold*.

(10.20) Definition. *A 3-dimensional pseudomanifold is a connected 3-complex K constructed by identifying faces, edges, and vertices of a set of solid polyhedra so that the following conditions are met:*

(1) *If $\sigma \in K$, then $\sigma < \tau$ for some 3-cell $\tau \in K$.*
(2) *Every 2-cell K is a face of exactly two 3-cells in K.*
(3) *If σ_1 and σ_n are 3-cells in K, then there is a sequence of 3-cells σ_1, $\sigma_2, \ldots, \sigma_n$ such that σ_i and σ_{i+1} intersect in a 2-cell.*

Thus, a pseudomanifold is constructed in analogy with a surface satisfying Definition 4.11, but in spite of the best intentions may not be a manifold, as in Example 10.19.

▷ **Exercise 10.7.** Show that the space M obtained from the cube C by gluing points of the form $(1, y, z)$ to $(-1, -y, z)$, $(x, 1, z)$ to $(x, -1, -z)$, and $(x, y, 1)$ to $(x, y, -1)$ is a 3-manifold.

3-manifolds are notoriously difficult to analyze. Many of the techniques we have developed for use on surfaces fail to yield results.

(10.21) Theorem. *If M is a compact connected 3-manifold, then $\chi(M) = 0$.*

Proof. M can be made from a set of solid polyhedra by identifying faces, edges, and vertices. Let V, E, F, and S denote the numbers of vertices, edges, faces, and solid polyhedra in M, so $\chi(M) = V - E + F - S$. Let V', E', F', and S' denote the numbers of vertices, etc., in the set of polyhedra before gluing them together. As in Example 10.18, each vertex $P_1, P_2, \ldots, P_V$ in M has a neighborhood topologically equivalent to a solid ball, with boundary a sphere. Thus, for each vertex we get a complex K_i with $|K_i| = \mathbb{S}^2$ and K_i has v_i vertices, e_i edges, and f_i faces, for $i = 1, 2, \ldots, V$. In Figure 10.20, we studied such a complex for the vertex of the 3-torus. By Theorem 5.10, $v_i - e_i + f_i = 2$, since $\chi(K_i) = \chi(\mathbb{S}^2) = 2$

Each edge in M connects two vertices, and so each edge intersects two of the K_i complexes, as in Figure 10.22. The vertices of the K_i complexes occur precisely at the points where the edges of M intersect K_i. Therefore,

$$2E = v_1 + v_2 + \cdots + v_V$$

Fig. 10.22. Two vertices and their associated complexes

Each face of the K_i complexes corresponds to one of the vertices of M before gluing as in Figure 10.23.

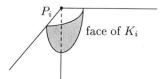

Fig. 10.23. Each face of K_i corresponds to a vertex before gluing

Therefore,

$$V' = f_1 + f_2 + \cdots + f_V$$

Let v'_i and e'_i denote the numbers of vertices and edges in the complex K_i before assembly. Before gluing, K_i is a bunch of polygons, so $v'_i = e'_i$. The edges will be identified in pairs to form K_i, so $v'_i = e'_i = 2e_i$.

▷ **Exercise 10.8.**   Prove the following:

$$2E' = v'_1 + v'_2 + \cdots + v'_V.$$
$$2E' = 2e_1 + 2e_2 + \cdots + 2e_V$$
$$E' = e_1 + e_2 + \cdots + e_V$$

It follows from Exercise 10.8 and the arguments above that

$$
\begin{aligned}
2E - E' + V' &= (v_1 + v_2 + \cdots + v_V) - (e_1 + e_2 + \cdots + e_V) \\
&\quad + (f_1 + f_2 + \cdots + f_V) \\
&= (v_1 - e_1 + f_1) + (v_2 - e_2 + f_2) + \cdots + (v_V - e_V + f_V) \\
&= 2 + 2 + \cdots + 2 = 2V
\end{aligned}
$$

It is obvious that $S = S'$. Since M is constructed by identifying the faces in pairs, $F' = 2F$. Each solid is a polyhedron, whose boundary is topologically equivalent to a sphere, so

$$V' - E' + F' = 2S'$$
$$V' - E' + 2F = 2S$$

Therefore,

$$\chi(M) = V - E + F - S = V - E + F - \left(\frac{V'}{2} - \frac{E'}{2} + F\right)$$
$$= V - E - \frac{1}{2}(V' - E')$$
$$= V - E - \frac{1}{2}(2V - 2E)$$
$$= 0$$

$\square$

Thus, the euler characteristic cannot differentiate between 3-manifolds. The converse of Theorem 10.21 is also true, but we will not prove it.

(10.22) Theorem. *Let M be a finite 3-complex constructed by identifying faces, edges, and vertices of a set on polyhedra. Then M is a 3-manifold if and only if $\chi(M) = 0$.*

In analogy with Definition 4.7, we define:

(10.23) Definition. *A 3-manifold is non-orientable if it contains a Klein bottle.*

▷ **Exercise 10.9.** Determine whether the following spaces formed by identifying the sides of the cube C are 3-manifolds, where

$$C = \{(x, y, z) : -1 \le x \le 1, \ -1 \le y \le 1, \ -1 \le z \le 1\}$$

For those which are manifolds, determine whether they are orientable.

(1)
$$\begin{cases} (1, y, z) \sim (-1, z, y) \\ (x, 1, z) \sim (z, -1, x) \\ (x, y, -1) \sim (y, x, -1) \end{cases}$$

(2)
$$\begin{cases} (1, y, z) \sim (-1, -y, -z) \\ (x, 1, z) \sim (x, -1, z) \\ (x, y, -1) \sim (x, y, -1) \end{cases}$$

$$(3) \qquad \begin{cases} (1, y, z) \sim (-1, -y, z) \\ (x, 1, z) \sim (x, -1, z) \\ (x, y, -1) \sim (x, y, -1) \end{cases}$$

$$(4) \qquad \begin{cases} (1, y, z) \sim (-1, -y, -z) \\ (x, 1, z) \sim (-x, -1, -z) \\ (x, y, -1) \sim (-x, -y, -1) \end{cases}$$

Chapter 11

Topology and calculus

"So far as the theories of mathematics are about reality, they are not certain; so far as they are certain, they are not about reality"

Albert Einstein

11.1 Vector fields and differential equations in $\mathbb{R}^n$

(11.1) Definition. *Let $A \subseteq \mathbb{R}^n$. A vector field $\mathbf{v}$ on A is a continuous function $\mathbf{v} : A \longrightarrow \mathbb{R}^n$. Thus, $\mathbf{v}$ associates to every point $\mathbf{x} \in A$ a vector $\mathbf{v}(\mathbf{x})$ in $\mathbb{R}^n$.*

A vector field is pictured as a vector attached at each point of the set, as in the examples in Figures 11.1 and 11.2. Thus, one can think of a vector field as a set with hair. Vector fields are common in nature: as force fields in gravity or electromagnetics, or as velocity vectors for fluids, such as the wind or water.

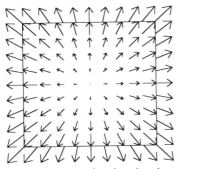

Fig. 11.1. $\mathbf{v}(x, y) = (x, y)$ **Fig. 11.2.** $\mathbf{v}(x, y) = (1 - x^2, 2xy)$

(11.2) Definition. *A point where $\mathbf{v}(\mathbf{x}) = 0$ is called a critical point for the vector field. An isolated critical point $\mathbf{x}$ is a critical point such that there are no other critical points in some neighborhood of $\mathbf{x}$.*

We consider only vector fields with isolated critical points in this text.

▷ **Exercise 11.1.** Sketch the following vector fields on $\mathbb{R}^2$, and identify all critical points.

(1) $\mathbf{v}(x, y) = (1, y)$
(2) $\mathbf{v}(x, y) = (x, -y)$
(3) $\mathbf{v}(x, y) = (y, x)$
(4) $\mathbf{v}(x, y) = (y, -x)$
(5) $\mathbf{v}(x, y) = (4x, y)$
(6) $\mathbf{v}(x, y) = (x^2 - 1, y)$
(7) $\mathbf{v}(x, y) = (y^2 - x^2, 2xy)$
(8) $\mathbf{v}(x, y) = (x, x^2 - y^2)$

A vector field $\mathbf{v}$ on $\mathbb{R}^n$ can be written as

$$\mathbf{v}(\mathbf{x}) = (v_1(\mathbf{x}), v_2(\mathbf{x}), \ldots, v_n(\mathbf{x}))$$

where $v_1, v_2, \ldots, v_n : \mathbb{R}^n \longrightarrow \mathbb{R}$ are the coordinate functions of $\mathbf{v}$ and $\mathbf{x} = (x_1, x_2, \ldots, x_n) \in \mathbb{R}^n$. The vector field can be interpreted as the system of differential equations:

$$\frac{dx_1}{dt} = v_1(\mathbf{x}), \quad \frac{dx_2}{dt} = v_2(\mathbf{x}), \quad \ldots, \quad \frac{dx_n}{dt} = v_n(\mathbf{x})$$

If one considers a vector field as a system of differential equations, then the set of solutions forms a family of paths on $\mathbb{R}^n$, with the vector $\mathbf{v}(\mathbf{x})$ tangent to the path through $\mathbf{x}$ for every $\mathbf{x} \in \mathbb{R}^n$.

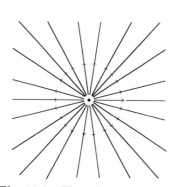

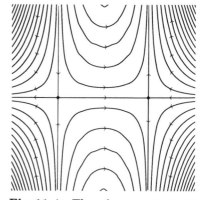

Fig. 11.3. The phase portrait of $\mathbf{v}(x, y) = (x, y)$

Fig. 11.4. The phase portrait of $\mathbf{v}(x, y) = (x^2 - 1, 2xy)$

By the existence and uniqueness theorems for the solutions of differential equations (see any text on differential equations), there is exactly one such path through each point $\mathbf{x}$ where $\mathbf{v}(\mathbf{x}) \neq 0$, but at the critical points of $\mathbf{v}$ the paths may split or join. These paths with the natural directions given by the vector field form a *phase portrait* of the vector field. The paths are called *flows*. Figures 11.3 and 11.4 give the phase portraits of the vector fields of Figures 11.1 and 11.2.

▷ **Exercise 11.2.**   Sketch the phase portraits for the vector fields of Exercise 11.1.

Note that the resulting phase portraits for vector fields on $\mathbb{R}^2$ look a lot like topographical maps or weather maps. A topographical map gives the contour lines of the height function over the area mapped. Contour lines are the set of points where the height is a chosen constant. If $z = F(x,y)$ defines a function $F : \mathbb{R}^2 \longrightarrow \mathbb{R}$, then the contour lines of the graph of F are given by the equation $F(x,y) = $ constant, and then along one of these contour lines we have

$$\frac{\partial F}{\partial x}dx + \frac{\partial F}{\partial y}dy = 0$$

Therefore, along these contour lines,

$$\frac{dy}{dx} = -\frac{\frac{\partial F}{\partial x}}{\frac{\partial F}{\partial y}}$$

Thus the contour lines of the graph of $z = F(x,y)$ are the flow lines of the vector field defined by

$$\mathbf{v}(x,y) = \left(\frac{\partial F}{\partial y}, -\frac{\partial F}{\partial x}\right)$$

The vectors of the field are tangent to the flow lines.

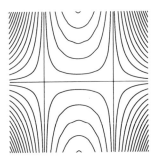

Fig. 11.5. A topographical map of $F(x,y) = y - yx^2$

For example, the vector field $\mathbf{v}(x,y) = (1 - x^2, 2xy)$ of Figure 11.2 is induced by the function $F(x,y) = y - yx^2$. Compare Figures 11.4 and 11.5. The contour lines or topographical map is derived by setting F equal to a constant. For example, if $y - yx^2 = 0$, then $y = 0$ or $x = \pm 1$. Thus, the lines $y = 0$ and $x = \pm 1$ are the 0-level set. If $y - yx^2 = k$, then $y = \frac{k}{1-x^2}$, and the contour lines are shown in Figure 11.5.

For a vector field whose component functions have continuous derivatives it is easy to check whether it is possible to find a function F which induces the vector field, so that $\mathbf{v}(x,y) = (\frac{\partial F}{\partial y}, -\frac{\partial F}{\partial x})$. If the function F exists, then its second derivatives will be continuous and so

$$\frac{\partial^2 F}{\partial x \partial y} = \frac{\partial^2 F}{\partial y \partial x}$$

If $\mathbf{v}(x,y) = (v_1(x,y), v_2(x,y)) = (\frac{\partial F}{\partial y}, -\frac{\partial F}{\partial x})$, then $\frac{\partial v_1}{\partial x} = -\frac{\partial v_2}{\partial y}$. Thus, if $\frac{\partial v_1}{\partial x} \neq -\frac{\partial v_2}{\partial y}$, then $\mathbf{v}$ is not induced by a function F. The vector field $\mathbf{v}(x,y) = (x,y)$ of Figure 11.1 is not induced by a function since $\frac{\partial v_1}{\partial x} = 1$ and $-1 = -\frac{\partial v_2}{\partial y}$.

▷ **Exercise 11.3.**   Determine which of the vector fields of Exercise 11.1 are induced by a function $F : \mathbb{R}^2 \longrightarrow \mathbb{R}$, and find the function F where it exists.

Let us first worry about the behavior of vector fields on the plane. The vector field flows smoothly and predictably except at the critical points. Let $\mathbf{x} \in \mathbb{R}^2$ be an isolated critical point of $\mathbf{v}$ and let D be a disc neighborhood of $\mathbf{x}$ so that there are no other critical points in D or on its boundary. The boundary of D is a circle $C = \partial(D)$. The vector field $\mathbf{v}$ attaches a non-zero vector at each point of the circle C. Pretend you are walking around C in the counterclockwise direction and carrying a divining rod which magically points in the direction of the vector field. When you return to your starting point, the divining rod also returns to its original position, so during the trip, the divining rod must have made some integer number of counterclockwise revolutions, where we count a clockwise revolution as -1. This number is called the *winding number* $w_{\mathbf{v}}(C)$ of the vector field on the cycle C. The curve C need not be a circle but can be any Jordan curve, i.e., any curve in $\mathbb{R}^2$ topologically equivalent to the circle, which will enclose a region homeomorphic to a disc by the Schoenflies Theorem 10.16. Examples are shown in Figures 11.6 and 11.7.

Another way of computing the winding number is to normalize the vector field on the circle C so that every vector has length 1, i.e., consider $f(\mathbf{x}) = \frac{\mathbf{v}(\mathbf{x})}{\|\mathbf{v}(\mathbf{x})\|}$. If one imagines a vector field as a hairy surface; you have just given it a haircut, so that every hair has the same length. Since C can be thought of as the circle $\mathbb{S}^1$, and the set of all vectors of length 1 in $\mathbb{R}^2$ is also a circle, f can be thought of as a function $f : \mathbb{S}^1 \longrightarrow \mathbb{S}^1$. The induced function $f_1 : H_1(\mathbb{S}^1) \simeq \mathbb{Z} \longrightarrow H_1(\mathbb{S}^1) \simeq \mathbb{Z}$ must be multiplication by some integer, since it is a homomorphism. This integer is the winding number of the vector field v on C. This idea generalizes to higher dimensions, when it is called the *degree* of the function f. By Example 6.17, $H_n(\mathbb{S}^n) \simeq \mathbb{Z}$.

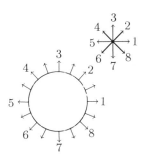

Fig. 11.6. $w_v(C) = 1$

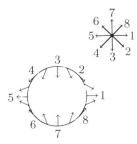

Fig. 11.7. $w_v(C) = -1$

(11.3) Definition. *Let $f : S^n \longrightarrow S^n$ be a continuous function. Then f induces a homomorphism $f_n : H_n(S^n) \simeq \mathbb{Z} \longrightarrow H_n(S^n) \simeq \mathbb{Z}$, so f_n is multiplication by some integer k. Define the degree of f as $\deg(f) = k$.*

▷ **Exercise 11.4.** Prove that if $f, g : S^n \longrightarrow S^n$ are homotopic continuous functions [see Definition 9.1 and Theorem 9.2], then $\deg(f) = \deg(g)$.

▷ **Exercise 11.5.** Prove that if $f, g : S^n \longrightarrow S^n$ are continuous functions, then the composition satisfies $\deg(f \circ g) = \deg(f) \cdot \deg(g)$.

▷ **Exercise 11.6.** Prove that $\deg(1_{S^n}) = 1$ for the identity function on the n-sphere.

▷ **Exercise 11.7.** Prove that if $f : S^n \longrightarrow S^n$ is a homeomorphism, then $\deg(f) = \pm 1$.

▷ **Exercise 11.8.** The trivial or constant function $f : S^n \longrightarrow S^n$ is defined by choosing a point $\mathbf{z} \in S^n$ and letting $f(\mathbf{x}) = \mathbf{z}$ for every $\mathbf{x} \in S^n$. Prove that $\deg(f) = 0$.

(11.4) Theorem. *Consider $S^n \subseteq \mathbb{R}^{n+1}$ as the set of points*

$$S^n = \{\mathbf{x} = (x_1, x_2, \ldots, x_{n+1}) : \|\mathbf{x}\| = 1\}$$

The function $f : S^n \longrightarrow S^n$ defined by

$$f(x_1, x_2, \ldots, x_{n+1}) = (-x_1, x_2, \ldots, x_{n+1})$$

has degree $\deg(f) = -1$.

Proof. We choose a simplicial complex to represent the sphere, and compute the degree of f for this complex. Note that the n-sphere can be considered as the boundary of an $(n+1)$-simplex $\sigma = \langle v_0, v_1, v_2, \ldots, v_{n+1} \rangle$ where $v_0, \ldots, v_{n+1} \in \mathbb{R}^{n+1}$ are the vertices of the simplex. Let

$$v_0 = (1, 0, 0, \ldots, 0) \quad \text{and} \quad v_1 = (-1, 0, 0, \ldots, 0)$$

Place the remaining vertices $v_2, \ldots, v_{n+1}$ on the n-dimensional plane consisting of points of the form $\{(0, x_2, \ldots, x_{n+1})\} \subseteq \mathbb{R}^{n+1}$ (Fig. 11.8).

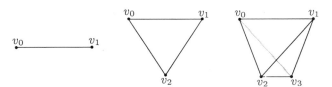

Fig. 11.8. The simplex σ for dimensions $1, 2$, and 3

Note that $f(v_0) = v_1$, $f(v_1) = v_0$, and $f(v_i) = v_i$ for $i \neq 0, 1$. Thus, by Lemma 7.6,

$$
\begin{aligned}
f(\mathbb{S}^n) &= f(\partial(\sigma)) \\
&= f(\partial(\langle v_0, v_1, v_2, \ldots, v_{n+1} \rangle)) \\
&= \partial(f(\langle v_0, v_1, v_2, \ldots, v_{n+1} \rangle)) \\
&= \partial(\langle v_1, v_0, v_2, \ldots, v_{n+1} \rangle) \\
&= \partial(-\langle v_0, v_1, v_2, \ldots, v_{n+1} \rangle) \\
&= -\partial(\sigma)
\end{aligned}
$$

Therefore, $deg(f) = -1$. $\qquad\qquad\qquad\qquad\qquad\qquad\qquad\qquad\qquad \square$

The function f of Theorem 11.4 is the reflection of $\mathbb{R}^{n+1}$ about the subspace $\{(0, x_2, \ldots, x_{n+1})\} \simeq \mathbb{R}^n \subseteq \mathbb{R}^{n+1}$, and reverses the orientation on the simplex σ.

▷ **Exercise 11.9.** Prove that the antipodal function $f : \mathbb{S}^n \longrightarrow \mathbb{S}^n$ defined by $f(\mathbf{x}) = -\mathbf{x}$ has degree $deg(f) = (-1)^{n+1}$.

We generalize the definition of the winding number of a vector field along a Jordan curve to dimensions greater than 2. If $\mathbf{v}$ is a vector field on $\mathbb{R}^{n+1}$, then $\frac{\mathbf{v}}{\|\mathbf{v}\|}$ is defined on the set of points where $\mathbf{v}$ is not zero and is a continuous function to the n-sphere.

(11.5) Definition. *Let $D \subseteq \mathbb{R}^{n+1}$ be a set topologically equivalent to the n-dimensional ball with boundary $C = \partial(D)$ and let $f : \mathbb{S}^n \longrightarrow C$ be the ensuing homeomorphism. Let $\mathbf{v}$ be a vector field defined on D with no zeroes on C and define $\overline{\mathbf{v}} : C \longrightarrow \mathbb{S}^n$ by*

$$\overline{\mathbf{v}}(\mathbf{x}) = \frac{\mathbf{v}(\mathbf{x})}{\|\mathbf{v}(\mathbf{x})\|}$$

The winding number of $\mathbf{v}$ on C is defined as $w_{\mathbf{v}}(C) = deg(\overline{\mathbf{v}} \circ f)$.

By Exercise 11.7 and Theorem 11.4, the choice of the homeomorphism f in Definition 11.5 essentially chooses an orientation for C.

▷ **Exercise 11.10.** Let $\mathbf{v}$ be a vector field defined on $\mathbb{R}^{n+1}$ and consider a homeomorphism $g : \mathbb{R}^{n+1} \longrightarrow \mathbb{R}^{n+1}$. Note that $\mathbf{v}' = \mathbf{v} \circ g^{-1}$ defines another vector field. If $C \subseteq \mathbb{R}^{n+1}$ is homeomorphic to $\mathbb{S}^n$ such that $\mathbf{v}$ has no zeroes on C and C bounds a region homeomorphic to the $(n+1)$-ball $\mathbb{B}^{n+1}$, then show that $w_{\mathbf{v}}(C) = w_{\mathbf{v}'}(g(C))$.

Exercise 11.8 and Definition 11.5 imply that the winding number of the constant vector field is zero when computed along any Jordan curve (Fig. 11.9).

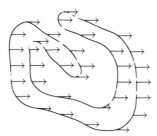

Fig. 11.9. The constant vector field $\mathbf{v}$ with $w_{\mathbf{v}}(C) = 0$

(11.6) Theorem. *If D is homeomorphic to the n-ball with the boundary C, and the vector field $\mathbf{v}$ is never zero on D, then $w_{\mathbf{v}}(C) = 0$.*

Proof. Since $\mathbf{v}$ is never zero on D, the function $\overline{\mathbf{v}} : D \longrightarrow \mathbb{S}^n$ defined by $\overline{\mathbf{v}} = \frac{\mathbf{v}}{\|\mathbf{v}\|}$ is continuous on all of D. Thus, the composition $\overline{\mathbf{v}} \circ f$ can be considered as a continuous function on the n-ball $\mathbb{B}^n$. Since B^n deformation retracts to a point, there is a homotopy from $\overline{\mathbf{v}} \circ f$ to the constant function. By Exercises 11.4 and 11.8, $w_{\mathbf{v}}(C) = deg(\overline{\mathbf{v}} \circ f) = 0$. □

If the vector field has zeroes inside the set D, then the proof above fails, since in this case the function $\overline{\mathbf{v}} \circ f$ is not defined on the n-ball. It thus seems that the critical points must be the only things that affect the winding number.

(11.7) Lemma. *If* v *is a vector field on* $\mathbb{R}^2$ *and* C_1 *and* C_2 *are 1-cycles such that* v *is never zero on* C_1 *and* C_2. *Then* $w_v(C_1 + C_2) = w_v(C_1) + w_v(C_2)$.

▷ **Exercise 11.11.** Prove Lemma 11.7 for dimension 2, considering separately the case where C_1 and C_2 intersect at either a point or not at all, and the case where C_1 and C_2 have an edge in common.

(11.8) Definition. *Let* v *be a vector field on with isolated critical point* x. *Let* D *be a disc neighborhood of* x *with boundary* C *such that there are no other critical points in* D *or on* C. *The index of* x *is given by* $I_v(x) = w_v(C)$.

Note that the index is well defined, since if D_1 and D_2 are two different discs about x, and $C_i = \partial(D_i)$, then the region between C_1 and C_2 can be divided into two regions A and B which are each homeomorphic to a disc, as in Figure 11.10. Note that $\partial(A) = a + e - c - f$ and $\partial(B) = b + f - d - e$, and the curves $C_1 = a + b$ and $C_2 = c + d$.

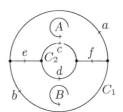

Fig. 11.10. $w_v(C_1) = w_v(C_2)$

By Theorem 11.6, we know that

$$w_v(\partial(A)) = w_v(a + e - c - f) = 0$$

$$w_v(\partial(B)) = w_v(b + f - d - e) = 0$$

Therefore, using Lemma 11.7, we have

$$w_v(C_1) - w_v(C_2) = w_v(a + b - c - d)$$
$$= w_v((a + e - c - f) + (b + f - d - e))$$
$$= w_v(\partial(D_1)) + w_v(\partial(D_2))$$
$$= 0$$

and so $w_v(C_1) = w_v(C_2)$.

Some common types of isolated critical points are illustrated in Figure 11.11. Note that a slight perturbation of a center will change it into a focus, where the flow lines spiral around forever, but never reach the critical point.

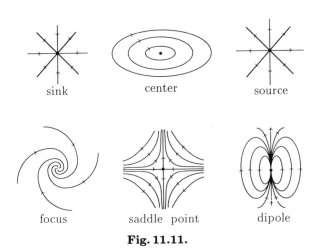

sink   center   source

focus   saddle point   dipole

Fig. 11.11.

Except for the center and the focus, notice that each of the examples of Figure 11.11 can be broken down into three types of sectors, as illustrated in Figure 11.12.

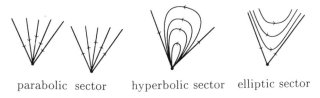

parabolic sector   hyperbolic sector   elliptic sector

Fig. 11.12. A parabolic sector, an elliptic sector, and a hyperbolic sector

(11.9) Poincaré-Bendixson Theorem. *If* **v** *is a vector field with isolated critical point* **x** *which is not a center or a focus, let e denote the number of elliptical sectors around* **x**, *and h the number of hyperbolic sectors around* **x**. *Then*

$$I_{\mathbf{v}}(\mathbf{x}) = 1 + \frac{e - h}{2}$$

The winding number can be though of as $\frac{1}{2\pi}$ times the total change in the angle swept out by the vector $\mathbf{v}(\mathbf{x})$ as $\mathbf{x}$ travels counterclockwise around a circle centered at the critical point.

▷ **Exercise 11.12.** Prove Theorem 11.9, by using the following outline:

(1) First note that the sum of the angle measures of all the sectors around $\mathbf{x}$ is 2π.

(2) Prove that if $\mathbf{v}$ has a parabolic sector of measure θ radians, then the change in the angle of the vector field across the sector is θ.

(3) Prove that if $\mathbf{v}$ has a elliptical sector of measure θ radians, then the change in the angle of the vector field across the sector is $\theta + \pi$.

(4) Prove that if $\mathbf{v}$ has a hyperbolic sector of measure θ radians, then the change in the angle of the vector field across the sector is $\theta - \pi$.

(5) Prove Theorem 11.9 for a vector field $\mathbf{v}$ and an isolated critical point $\mathbf{x}$ which has a finite number of sectors of these three types.

▷ **Exercise 11.13.** Find the indices of the critical points of Figure 11.13.

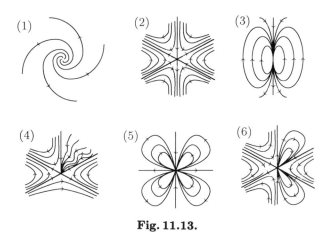

Fig. 11.13.

▷ **Exercise 11.14.** Draw isolated critical points of index -2, -1, 0, 1, 2.

(11.10) Poincaré Index Theorem for the Disc. *Let D^n be homeomorphic to an n-ball with $C^{n-1} = \partial(D)$, and let $\mathbf{v}$ be a continuous vector field with only isolated critical points $\mathbf{x}_1, \mathbf{x}_2, \dots, \mathbf{x}_k \in D$. Then*

$$w_{\mathbf{v}}(C) = \sum_{i=1}^{k} I_{\mathbf{v}}(\mathbf{x}_i)$$

Proof. Since D is compact, $\mathbf{v}$ can have only a finite number of isolated critical points. For each critical point $\mathbf{x}_i$ find a n-dimensional closed disc neighborhood D_i so that $\mathbf{x}_i$ is the only critical point in D_i or its boundary $C_i = \partial(D_i)$. These discs can also be chosen so that these discs are disjoint. Then $w_{\mathbf{v}}(C_i) = I_{\mathbf{v}}(\mathbf{x}_i)$. Connect each C_i to the outer circle C by an edge a_i, so that the edges a_i do not intersect each other or the discs D_i. In dimension 2, note that the set $D' = Int(D) - \cup_{i=1}^k a_i - \cup_{i=1}^k D_i$ is topologically equivalent to an open disc (see Fig. 11.14). In higher dimensions, it is necessary to thicken up the chosen edges into solid tubes, and consider the set $D' = D - \cup_{i=1}^k \partial(\text{tube}_i) - \cup_{i=1}^k D_i$ which is homeomorphic to the n-ball, as in Figure 11.15.

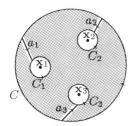

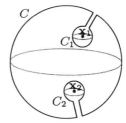

Fig. 11.14. D' is homeomorphic to a disc

Fig. 11.15. D' is homeomorphic to the 3-ball

Since the vector field $\mathbf{v}$ has no zeroes in D', the winding number is zero on $\partial(D')$. The boundary $\partial(D')$ consists of C, the C_i's, and the edges a_i or the associated tubes. Since the edges and the tubes were chosen to avoid all the critical points, we have $w_{\mathbf{v}}(C_i + a_i - a_i) = w_{\mathbf{v}}(C_i)$ in dimension 2 by Lemma 11.7, and in higher dimensions it can also be shown that $w_{\mathbf{v}}(C_i \cup \text{tube}_i) = w_{\mathbf{v}}(C_i)$. Therefore,

$$0 = w_{\mathbf{v}}(\partial(D')) = w_{\mathbf{v}}(C) - \sum_{i=1}^k w_{\mathbf{v}}(C_i)$$

$$= w_{\mathbf{v}}(C) - \sum_{i=1}^k I_{\mathbf{v}}(\mathbf{x}_i)$$

and, thus, $w_{\mathbf{v}}(C) = \sum_{i=1}^n I_{\mathbf{v}}(\mathbf{x}_i)$.   □

As an application, consider a desert island and the height above sea level defined on that island. This defines a continuous function h defined on the set I (for island), which in turn induces a vector field $\mathbf{v} = (\frac{\partial h}{\partial y}, -\frac{\partial h}{\partial x})$. Note that the critical points of the vector field are the critical points of the function h, i.e., points with a horizontal tangent plane. The island I has coastline C homeomorphic to a circle, and the vector field $\mathbf{v}$ gives an orientation (clockwise or counterclockwise) on C. The vectors $\mathbf{v}$ along C are tangent to C.

▷ **Exercise 11.15.**   Show that $w_{\mathbf{v}}(C) = 1$.

The most common critical points on such an island are mountain peaks and pits (circular valleys) which are centers for the vector field and so have index $+1$, and saddle points or mountain passes, as in Figure 11.16, which have index -1.

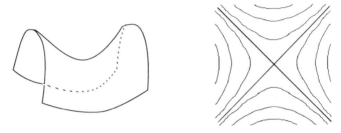

Fig. 11.16. A saddle point and its topographical map

It follows from the Poincaré Index Theorem 11.10 that if there are no other types of critical points, then

$$\text{peaks} + \text{pits} - \text{passes} = 1$$

▷ **Exercise 11.16.**   Draw topographical maps of imaginary islands with:
(1) two peaks and one pass
(2) one peak, one pit, and one pass

▷ **Exercise 11.17.**   Consider the case of an island with several lakes in the middle. Show that peaks + pits − passes + lakes = 1.

▷ **Exercise 11.18.**   Draw a topological map of an imaginary island with two lakes and two passes and whatever else you need to make it work.

11.2 Differentiable manifolds

In order to be able to do calculus on manifolds, we must impose additional structure on the manifolds and the functions. This task is easier if we assume that all manifolds are subsets of some $\mathbb{R}^N$.

(11.11) Definition. *A function $f : \mathbb{R}^n \longrightarrow \mathbb{R}^m$ is smooth if f has continuous partial derivatives of all orders.*

If $f : \mathbb{R}^n \longrightarrow \mathbb{R}^m$ is a smooth function, then it can be approximated at any point $\mathbf{x} \in \mathbb{R}^n$ by its differential $df_{\mathbf{x}} : \mathbb{R}^n \longrightarrow \mathbb{R}^m$ where $df_{\mathbf{x}}$ is a linear transformation. For functions $f : \mathbb{R} \longrightarrow \mathbb{R}$ this is just the standard approximation of the curve by its tangent line. In higher dimensions, if $\mathbf{x} = (x_1, x_2, \ldots, x_n)$ and $f(\mathbf{x}) = (f_1(\mathbf{x}), f_2(\mathbf{x}), \ldots, f_m(\mathbf{x}))$, then df is usually written as the m by n matrix

$$df_{\mathbf{x}} = \left[\frac{\partial f_i}{\partial x_j}(\mathbf{x}) \right]$$

(11.12) Definition. *Let $A \subseteq \mathbb{R}^N$. A function $f : A \longrightarrow B \subseteq \mathbb{R}^k$ is a diffeomorphism if f is a homeomorphism and both f and its inverse are smooth.*

Fig. 11.17. These are homeomorphic, but not diffeomorphic

(11.13) Definition. *A smooth n-manifold is an n-manifold $M \subseteq \mathbb{R}^N$ such that every point $\mathbf{x} \in M$ has a disc neighborhood U relative to M with a diffeomorphism $f : D^n \longrightarrow U$, where D^n is a disc in $\mathbb{R}^n$.*

Figure 11.17 gives some examples of spaces which are homeomorphic, as studied in Chapter 4, but not diffeomorphic. If a manifold M is smooth, then the neighborhood $U = f(D^n)$ about the point $\mathbf{x}$ can be approximated by $df_{\mathbf{x}}(D^n)$. Since $df_{\mathbf{x}}$ is a linear transformation and D^n is diffeomorphic to $\mathbb{R}^n$, we approximate the neighborhood U by the tangent space given by $T_{\mathbf{x}}M = df_{\mathbf{x}}(\mathbb{R}^n) \subseteq \mathbb{R}^N$. See Figure 11.18.

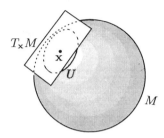

Fig. 11.18. The neighborhood U of $\mathbf{x}$ and the tangent plane $T_{\mathbf{x}}M$

The cube is not a smooth manifold, since there is no tangent plane at the corners or at the edges. Smooth manifolds are gently rounded; in fact, smooth.

The circle $\mathbb{S}^1 = \{(x,y) \in \mathbb{R}^2 : x^2 + y^2 = 1\}$ is a smooth manifold since at any point it has a well-defined tangent line. For example, near the point $(1,0) \in \mathbb{S}^1$, the function $f : \mathbb{R} \longrightarrow \mathbb{S}^1$ defined by $f(t) = (\cos(\pi t), \sin(\pi t))$ is a diffeomorphism from a small interval around $1 \in \mathbb{R}$ onto a small warped interval around $(1,0) \in \mathbb{S}^1$. The differential of f is given by the matrix

$$df = \begin{bmatrix} -\pi \sin(\pi t) \\ \pi \cos(\pi t) \end{bmatrix}$$

If we represent points on the circle as vectors $\mathbf{x} = (x,y)$, note that $df_{\mathbf{x}}(x,y) = \pi(-y, x)$. Recall the *dot product* defined in linear algebra for vectors $\mathbf{u} = (u_1, \dots, u_n)$ and $\mathbf{w} = (w_1, \dots, w_n)$ in $\mathbb{R}^n$ by

$$\mathbf{u} \cdot \mathbf{w} = u_1 w_1 + u_2 w_2 + \cdots + u_n w_n$$

The vectors $\mathbf{u}$ and $\mathbf{w}$ are perpendicular if and only if $\mathbf{u} \cdot \mathbf{w} = 0$. Since for the vector $\mathbf{x} = (x,y) \in \mathbb{S}^1$ we have $(x,y) \cdot df_{\mathbf{x}}(x,y) = x(-\pi y) + y(\pi x) = 0$, the tangent line given by $df_{\mathbf{x}}$ is perpendicular to the vector $\mathbf{x}$. The same is true for higher-dimensional spheres: The tangent space $T_{\mathbf{x}}\mathbb{S}^n$ is perpendicular to the vector $\mathbf{x} \in \mathbb{S}^n$.

In this text, we will more often appeal to pictures and intuition that to explicit computations of differentials and tangent spaces.

11.3 Vector fields on manifolds

Now we are equipped to study vector fields on manifolds.

(11.14) Definition. *Let $M \subseteq \mathbb{R}^N$ be a smooth n-manifold. A tangent vector field* **v** *on M is a continuous function* **v** $: M \longrightarrow \mathbb{R}^N$ *such that* **v**$(\mathbf{x}) \in T_{\mathbf{x}}M$ *for every* $\mathbf{x} \in M$. *Thus,* **v** *associates to every point* $\mathbf{x} \in M$ *a tangent vector* **v**$(\mathbf{x})$.

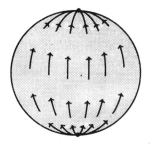

Fig. 11.19. A vector field on $\mathbb{S}^2$

Fig. 11.20. A vector field on $\mathbb{T}^2$

Note that the definitions of critical points, winding numbers, and indices from Chapter 11.1 can be extended to smooth manifolds since every point on a smooth manifold has a disc-like neighborhood. The vector field of Figure 11.19 has two critical points, each of index $+1$, whereas that of Figure 11.20 has no critical points.

(11.15) Theorem. *The n-sphere* $\mathbb{S}^n$ *has a non-zero vector field if and only if n is odd.*

Proof. If n is odd, then $n = 2k - 1$, and we can consider $\mathbb{S}^n \subseteq \mathbb{R}^{2k}$. Define a vector field **v** on $\mathbb{S}^n$ by

$$\mathbf{v}(x_1, x_2, \ldots, x_{2k-1}, x_{2k}) = (-x_2, x_1, \ldots, -x_{2k}, x_{2k-1})$$

By taking dot products, note that **v**$(\mathbf{x})$ is perpendicular to the vector **x**, so **v**$(\mathbf{x})$ lies in the tangent plane at **x** and satisfies Definition 11.14. The vector field **v** is never zero on $\mathbb{S}^n$.

If n is even and there is a non-zero vector field **v** on $\mathbb{S}^n$, then let $\overline{\mathbf{v}} = \frac{\mathbf{v}}{\|\mathbf{v}\|}$. Since the tangent space at **x** is perpendicular to the radial vector **x**, $\overline{\mathbf{v}}(\mathbf{x})$ cannot be a multiple of **x**. See Figure 11.21.

Define a function F for $(\mathbf{x}, t) \in \mathbb{S}^n \times I$ by

$$F(\mathbf{x}, t) = \cos(\pi t)\,\mathbf{x} + \sin(\pi t)\,\overline{\mathbf{v}}(\mathbf{x})$$

Note that

$$
\begin{aligned}
\|F(\mathbf{x}, t)\|^2 &= (F(\mathbf{x}, t)) \cdot (F(\mathbf{x}, t)) \\
&= (\cos(\pi t)\mathbf{x} + \sin(\pi t)\overline{\mathbf{v}}(\mathbf{x})) \cdot (\cos(\pi t)\mathbf{x} + \sin(\pi t)\overline{\mathbf{v}}(\mathbf{x})) \\
&= \cos^2(\pi t)\|\mathbf{x}\|^2 + 2\cos(\pi t)\sin(\pi t)(\mathbf{x} \cdot \overline{\mathbf{v}}(\mathbf{x})) + \sin^2(\pi t)\|\overline{\mathbf{v}}(\mathbf{x})\|^2
\end{aligned}
$$

Fig. 11.21. x and $\bar{v}(x)$

$$\|F(\mathbf{x}, t)\|^2 = \cos^2(\pi t) + \sin^2(\pi t)$$
$$= 1$$

since $\|\mathbf{x}\| = 1 = \|\bar{v}(\mathbf{x})\|$ and $\mathbf{x}$ and $\bar{v}(\mathbf{x})$ are perpendicular. Thus, $F(\mathbf{x}, t)$ can be thought of as a vector in $\mathbb{S}^n$, and F is a homotopy $F : \mathbb{S}^n \times I \longrightarrow \mathbb{S}^n$. The homotopy F is essentially the straight line from $\mathbf{x}$ to $\bar{v}(\mathbf{x})$ pushed out to the sphere. Note that $F(\mathbf{x}, 0) = \mathbf{x}$ and $F(\mathbf{x}, 1) = -\mathbf{x}$. Therefore, we have a homotopy from the identity function to the antipodal function f on $\mathbb{S}^n$. Since $deg(1_{\mathbb{S}^n}) = 1$ and $deg(f) = (-1)^{n+1}$, there is a contradiction when n is even. $\square$

The Poincaré Index Theorem 11.10 extends to vector fields on manifolds. As before, we assume that the vector field $\mathbf{v}$ has only isolated critical points.

▷ **Exercise 11.19.**   Prove that if $\mathbf{v}$ is a tangent vector field on a compact smooth manifold with only isolated critical points, then $\mathbf{v}$ has only a finite number of critical points.

(11.16) Lemma. *Any non-zero vector field on the circle $\mathbb{S}^1$ can be extended to a vector field on the 2-dimensional ball.*

▷ **Exercise 11.20.**   Prove Lemma 11.16.

(11.17) Poincaré Index Theorem. *Let $\mathbf{v}$ be a tangent vector field on a smooth surface S with only isolated critical points $\mathbf{x}_1, \mathbf{x}_2, \ldots, \mathbf{x}_k$. Then*

$$\sum_{i=1}^{k} I_{\mathbf{v}}(\mathbf{x}_i) = \chi(S)$$

Proof. From Theorem 4.14, all surfaces can be classified as either the sphere, the projective plane, a connected sum of tori, or a connected sum of projective planes. We prove each case separately.

The sphere: By Exercise 11.19, **v** has only a finite number of critical points, so we can choose a point **y** $\in$ $\mathbb{S}^2$ well away from all the critical points. By continuity, the vector field **v** is essentially constant on some very small neighborhood U about **y**. Let C denote the boundary of U. Note that C is also the boundary of $\mathbb{S}^2 - U$, and $\mathbb{S}^2 - U$ is diffeomorphic by stereographic projection to a disc.

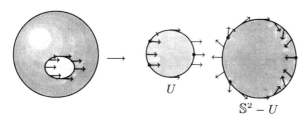

Fig. 11.22. The vector field **v** on $\mathbb{S}^2$, on U, and on $\mathbb{S}^2 - U$

The winding number of **v** must now be computed taking into account not only the curve C, but also which region surrounded by C is being considered. From Figure 11.22, we compute $w_{\mathbf{v}}(\partial(U)) = 0$ and, similarly, $w_{\mathbf{v}}(\partial(\mathbb{S}^2 - U)) = 2$. By Theorem 11.10, $w_{\mathbf{v}}(\partial(\mathbb{S}^2 - U)) = \sum_{i=1}^{k} I_{\mathbf{v}}(\mathbf{x}_i)$. Since all the critical points lie in $\mathbb{S}^2 - U$, it follows that

$$\sum_{i=1}^{k} I_{\mathbf{v}}(\mathbf{x}_i) = 2 = \chi(\mathbb{S}^2)$$

The projective plane: Recall from Example 7.17 that the projective plane is the sphere with opposite points identified. Any point in $\mathbb{P}^2$ can be written as $p(\mathbf{x})$ for $\mathbf{x} \in \mathbb{S}^2$ where $p(\mathbf{x}) = p(-\mathbf{x})$. If **v** is a tangent vector field on $\mathbb{P}^2$, then define a vector field $\tilde{\mathbf{v}}$ on $\mathbb{S}^2$ by $\tilde{\mathbf{v}}(\mathbf{x}) = \mathbf{v}(p(\mathbf{x}))$. If $p(\mathbf{x}_1)$, $p(\mathbf{x}_2)$, $\ldots$, $p(\mathbf{x}_k)$ are the critical points of **v**, then $\mathbf{v}(p(\mathbf{x}_i)) = 0 = \tilde{\mathbf{v}}(\mathbf{x}_i) = \tilde{\mathbf{v}}(-\mathbf{x}_i)$. Thus, $\mathbf{x}_1$, $-\mathbf{x}_1$, $\mathbf{x}_2$, $-\mathbf{x}_2$, $\ldots$, $\mathbf{x}_k$, and $-\mathbf{x}_k$ are the critical points of $\tilde{\mathbf{v}}$. Since $p : \mathbb{S}^2 \longrightarrow \mathbb{P}^2$ is the double covering transformation, p is a local diffeomorphism, and, thus, by Exercise 11.10,

$$I_{\mathbf{v}}(p(\mathbf{x})) = I_{\tilde{\mathbf{v}}}(\mathbf{x}) = I_{\tilde{\mathbf{v}}}(-\mathbf{x})$$

for $i = 1, 2, \ldots, k$. By the Poincaré Index Theorem for the sphere,

$$2 = \sum_{i=1}^{k} (I_{\tilde{\mathbf{v}}}(\mathbf{x}_i) + I_{\tilde{\mathbf{v}}}(-\mathbf{x}_i))$$

$$2 = 2 \sum_{i=1}^{k} I_{\mathbf{v}}(p(\mathbf{x}_i))$$

$$1 = \sum_{i=1}^{k} I_{\mathbf{v}}(p(\mathbf{x}_i)) = \chi(\mathbb{P}^2)$$

The torus: The following proof is adapted from Sieradski's *An Introduction to Topology and Homotopy*. Let $\mathbf{v}$ be a tangent vector field on $\mathbb{T}^2$. Since $\mathbf{v}$ has only a finite number of isolated critical points, we can choose a loop a as in Figure 11.23 such that a does not pass near any of the critical points. Cut the torus open along a, and note that $\partial(\mathbb{T}^2 - \{a\})$ consists of two disjoint circles a_1 and a_2. Attach two discs D_1 and D_2 smoothly along each circle to get a new manifold $\mathbb{T}'$. This type of modification to the manifold is called *surgery*.

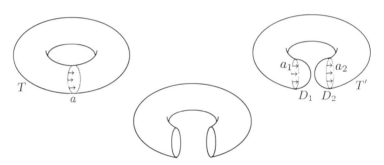

Fig. 11.23. Surgery along a

Note that $\mathbf{v}$ induces a non-zero vector field on $a_i = \partial(D_i)$. By Lemma 11.16, $\mathbf{v}$ can be extended to a vector field over the discs D_i. Thus, $\mathbf{v}$ induces a vector field $\mathbf{v}'$ on $\mathbb{T}'$ with the same critical points as $\mathbf{v}$ on the main portion of the torus, but with some new critical points on the attached discs.

Let $\mathbf{x}_1, \mathbf{x}_2, \ldots, \mathbf{x}_k$ be the critical points of $\mathbf{v}$, and $\mathbf{x}'_1, \ldots, \mathbf{x}'_m$ the critical points added by $\mathbf{v}'$ on $D_1 \cup D_2$. Note that $\mathbb{T}'$ is diffeomorphic to the sphere, and also that the discs D_1 and D_2 can be glued together along their boundary circles a_1 and a_2 to form a sphere. This is denoted by $D_1 \cup_a D_2$; see Figure 11.24.

In particular, the vector field $\mathbf{v}$ along the circle a matches up when the discs are glued together, so $\mathbf{v}'$ defines a vector field on $D_1^2 \cup_a D_2^2 = \mathbb{S}^2$.

$$2 = \chi(D_1^2 \cup_a D_2^2) = \sum_{i=1}^{m} (I_{\mathbf{v}'}(\mathbf{x}'_i))$$

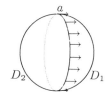

Fig. 11.24. $D_1 \cup_a D_2 = \mathbb{S}^2$

Therefore, by the index theorem for the sphere proved above,

$$2 = \chi(\mathbb{T}')$$
$$= \sum_{i=1}^{k} (I_{\mathbf{v}}(\mathbf{x}_i)) + \sum_{i=1}^{m} (I_{\mathbf{v}'}(\mathbf{x}_i'))$$
$$= \sum_{i=1}^{k} (I_{\mathbf{v}}(\mathbf{x}_i)) + 2$$
$$\chi(\mathbb{T}^2) = 0 = \sum_{i=1}^{k} (I_{\mathbf{v}}(\mathbf{x}_i))$$

The Klein bottle: Perform surgery on the handle of the Klein bottle as above. This can pictured on the planar diagram as in Figure 11.25.

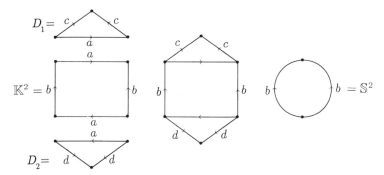

Fig. 11.25. Surgery on the Klein bottle

As above, we find that

$$\chi(\mathbb{K}^2) = 0 = \sum_{i=1}^{k} (I_{\mathbf{v}}(\mathbf{x}_i))$$

▷ **Exercise 11.21.** Prove the Poincaré Index Theorem for an n-handled torus, by doing surgery on each of the handles.

▷ **Exercise 11.22.** From Lemma 4.15, $3\mathbb{P}^2 = \mathbb{T}^2 \# \mathbb{P}^2$. Use this and surgery on the tori to prove the Poincaré Index Theorem for the connected sums of projective planes.

These exercises complete the proof of Theorem 11.17.   □

Given a triangulation on a surface, there is a canonical way of constructing a vector field corresponding to the simplicial structure. Place a sink in the middle of each triangle, a source at each vertex, and a hyperbolic critical point at the midpoint of each edge, as in Figure 11.26.

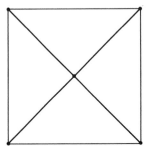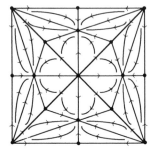

Fig. 11.26. Constructing a vector field for a triangulation

Since sinks and sources both have index +1, and saddle points have index −1, we have for this vector field

$$\sum_{i=1}^{k} I_{\mathbf{v}}(\mathbf{x}_i) = v - e + f = \chi(M)$$

(11.18) Corollary. *One cannot comb the hair on a coconut.*

(11.19) Corollary. *At any time, there is a point on the earth where the wind is not blowing.*

▷ **Exercise 11.23.** Draw a vector field on the torus with
   (1) one center and one saddle point
   (2) a dipole and two saddle points

▷ **Exercise 11.24.**   Draw a vector field on the sphere with

   (1) only one critical point
   (2) two centers
   (3) three critical points.

11.4 Integration on manifolds

In calculus, you computed a lot of integrals. Usually you integrated over sets which happened to be manifolds though, of course, at the time you did not know this. Topology forms a natural setting for calculus and can be used to simplify and clarify it. None of the theorems in this section are proved, since I am just trying to put things you already know into a new setting. Integrals on subsets of $\mathbb{R}^3$ can be divided into three types:

Line Integrals

$$\int_a^b f(x)\, dx \quad \text{or} \quad \int_C f(x)\, dx$$

where $[a, b]$ denotes an interval and C is a curve in $\mathbb{R}^n$.

Surface Integrals

$$\iint_R f(x, y)\, dxdy \quad \text{or} \quad \iint_S f\, dA$$

where R is a region in the plane and S is a 2-dimensional surface in $\mathbb{R}^n$.

Volume Integrals

$$\iiint_T f(x, y, z)\, dxdydz$$

where T is a 3-dimensional solid in $\mathbb{R}^n$.

The integrands (the things to be integrated) of these are called differential forms and can be written in standard form. We work with three variables, though fewer or more could be used.

(11.20) Definition. *A differential form ω on $\mathbb{R}^3$ is of one of the following forms:*

   (0) *0-forms: $\omega = f(x, y, z)$ where f is a differentiable function. 0-forms are evaluated at single points.*
   (1) *1-forms: $\omega = f(x, y, z)dx + g(x, y, z)dy + h(x, y, z)dz$, where f, g, h are differentiable functions. These 1-forms are integrated over intervals or over curves, i.e., over 1-manifolds.*

(2) *2-forms:* $\omega = f(x, y, z)dxdy + g(x, y, z)dzdx + h(x, y, z)dydz$ *where* f, g, h *are differentiable functions. These 2-forms are integrated over regions in the plane or over surfaces, i.e., over 2-manifolds.*

(3) *3-forms:* $\omega = f(x, y, z)dxdydz$ *where* f *is a differentiable function. 3-forms are integrated over solids, or 3-manifolds.*

(11.21) Definition. *If* $f(x, y, z)$ *is a differentiable function, the differential of* f *is defined by*

$$df = \frac{\partial f}{\partial x}dx + \frac{\partial f}{\partial y}dy + \frac{\partial f}{\partial z}dz$$

If ω is a 1-form, then $\omega = f dx + g dy + h dz$, or if we use variables x_1, x_2, x_3 instead of x, y, z and functions f_1, f_2, f_3 instead of f, g, h, then

$$\omega = f_1(x_1, x_2, x_3)dx_1 + f_2(x_1, x_2, x_3)dx_2 + f_3(x_1, x_2, x_3)dx_3$$

$$= \sum_{i=1}^{3} f_i(x_1, x_2, x_3)dx_i$$

The *differential* of the 1-form ω is the 2-form

$$d\omega = \sum_{i=1}^{3}\sum_{j=1}^{3} \frac{\partial f_i}{\partial x_j}dx_j dx_i$$

We are going to integrate over manifolds with orientation, so that if C is a curve from P to Q, then $-C$ is the same curve, but traveling from Q to P. Thus,

$$\int_{-C} f dx + g dy + h dz = -\int_{C} f dx + g dy + h dz$$

If one evaluates a double integral first in the x-direction, and then in the y-direction, one gets the opposite orientation as when one integrates y first, then x. Thus, the following theorem makes sense.

(11.22) Theorem. $dxdy = -dydx$

It follows that

$$dxdx = -dxdx$$

so that

$$dxdx = 0$$

Other rules for manipulating forms are included in the theorem below:

(11.23) Theorem. *Let* ω *and* η *be* k-*forms, and* λ *an* ℓ-*form.*

(1) $d(d\omega) = 0$

(2) $d(\omega + \eta) = d\omega + d\eta$

(3) $d(\omega\lambda) = (d\omega)\lambda + (-1)^k \omega(d\lambda)$

▷ **Exercise 11.25.**   Prove Theorem 11.23 for the cases

(1) $k = 0$ and $\ell = 1$
(2) $k = 1 = \ell$

If ω is a 1-form, then $\omega = f dx + g dy + h dz$ where $f, g, h : \mathbb{R}^3 \longrightarrow \mathbb{R}$ are differentiable functions.

$$d\omega = d(f dx + g dy + h dz)$$
$$= d(f dx) + d(g dy) + d(h dz) \quad \text{by (2) of Theorem 11.23.}$$
$$= (df)dx + f(d(dx)) + (dg)dy + g(d(dy)) + (dh)dz + h(d(dz))$$
$$= \left(\frac{\partial f}{\partial x} dx + \frac{\partial f}{\partial y} dy + \frac{\partial f}{\partial z} dz \right) dx + 0 + \left(\frac{\partial g}{\partial x} dx + \frac{\partial g}{\partial y} dy + \frac{\partial g}{\partial z} dz \right) dy + 0$$
$$\quad + \left(\frac{\partial h}{\partial x} dx + \frac{\partial h}{\partial y} dy + \frac{\partial h}{\partial z} dz \right) dz + 0$$

By Definition 11.21 and Theorem 11.22(3)

$$= \frac{\partial f}{\partial x} dx dx + \frac{\partial f}{\partial y} dy dx + \frac{\partial f}{\partial z} dz dx + \frac{\partial g}{\partial x} dx dy + \frac{\partial g}{\partial y} dy dy + \frac{\partial g}{\partial z} dz dy$$
$$\quad + \frac{\partial h}{\partial x} dx dz + \frac{\partial h}{\partial y} dy dz + \frac{\partial h}{\partial z} dz dz$$
$$= \left(\frac{\partial g}{\partial x} - \frac{\partial f}{\partial y} \right) dx dy + \left(\frac{\partial h}{\partial x} - \frac{\partial f}{\partial z} \right) dx dz + \left(\frac{\partial h}{\partial y} - \frac{\partial g}{\partial z} \right) dy dz$$

Thus, for a 1-form, $d\omega$ is a 2-form.

If ω is a 2-form, then ω can be written as

$$\omega = f(x, y, z) dx dy + g(x, y, z) dz dx + h(x, y, z) dy dz$$

and, therefore,

$$d\omega = (df)dx dy + f d(dx dy) + (dg)dz\, dx + g d(dy dz) + (dh)dy dz + h d(dy dz)$$
$$= \left[\left(\frac{\partial f}{\partial x} dx + \frac{\partial f}{\partial y} dy + \frac{\partial f}{\partial z} dz \right) dx dy + f(ddx dy - dx ddy) \right]$$
$$\quad + \left[\left(\frac{\partial g}{\partial x} dx + \frac{\partial g}{\partial y} dy + \frac{\partial g}{\partial z} dz \right) dz dx + g(ddz dx - dz ddx) \right]$$
$$\quad + \left[\left(\frac{\partial h}{\partial x} dx + \frac{\partial h}{\partial y} dy + \frac{\partial h}{\partial z} dz \right) dy dz + h(ddy dz - dy ddz) \right]$$
$$= \left(\frac{\partial f}{\partial z} \right) dz dx dy + \left(\frac{\partial g}{\partial y} \right) dy dz dx + \left(\frac{\partial h}{\partial x} \right) dx dy dz$$
$$= \left(\frac{\partial f}{\partial z} + \frac{\partial g}{\partial y} + \frac{\partial h}{\partial x} \right) dx dy dz$$

(11.24) Definition. *Let M be a smooth manifold. The group of k-forms is*

$$C^k(M) = \{k\text{-forms } \omega \text{ defined on } M\}$$

The differential operator defines a function $d : C^k(M) \longrightarrow C^{k+1}(M)$. It follows from Theorem 11.23 that this is a homomorphism. The differential works oppositely from the boundary homomorphism on k-chains defined in Chapter 6, where $\partial : C_k(M) \longrightarrow C_{k-1}(M)$. They are, however, related, as seen below:

(11.25) Fundamental Theorem of Calculus. *If f is a continuous function with antiderivative F (so $\frac{dF}{dx} = f$), then*

$$\int_a^b f(x)dx = F(b) - F(a)$$

This can be restated in terms of differential forms and chains by noting that $f(x)dx$ is a 1-form and F a 0-form with $dF = \frac{dF}{dx}dx = f(x)dx$. The interval $[a, b]$ is a 1-chain with $\partial[a, b] = b - a$.

(11.26) Stokes' Theorem or the Topologist's Fundamental Theorem of Calculus. *If ω is a k-form and C is a $(k + 1)$-chain, then*

$$\int_C d\omega = \int_{\partial C} \omega$$

Thus, the traditional Fundamental Theorem of Calculus is Theorem 11.26 for 0-forms:

$$\int_{[a,b]} dF = \int_{\partial[a,b]} F = F(b) - F(a)$$

where integrating a 0-form on a 0-chain is just evaluation of the function at the points.

(11.27) Green's Theorem. *Let R be a region in the plane*

$$\int_{\partial R} Pdx + Qdy = \iint_R \left(\frac{\partial Q}{\partial x} - \frac{\partial P}{\partial x} \right) dxdy$$

The traditional Green's Theorem is a restatement of Theorem 11.26 for a 1-form $\omega = Pdx + Qdy$. As above,

$$d\omega = \left(\frac{\partial Q}{\partial x} - \frac{\partial P}{\partial y} \right) dxdy$$

Stokes' Theorem (for 1-forms in three coordinates) and the Divergence Theorem (for 2-forms) of classical multivariable calculus are interpretations of Theorem 11.26 in different settings.

(11.28) Definition. *Let M be a smooth manifold. A k-form ω on M is called closed if $d\omega = 0$. The subgroup of closed forms is*

$$Z^k(M) = \{closed\ k\text{-}forms\ defined\ on\ M\}$$

Closed forms are analogous to cycles in homology. Closed forms are especially nice since if ω is a closed form and C and D are homologous chains, so that $C - D = \partial(E)$, then

$$\int_C \omega - \int_D \omega = \int_{C-D} \omega$$
$$= \int_{\partial(E)} \omega$$
$$= \int_E d(\omega)$$
$$= 0$$

and, thus,

$$\int_C \omega = \int_D \omega$$

To summarize: the integral of a closed form over a boundary chain is zero.

(11.29) Definition. *Let M be a smooth manifold. A k-form ω is exact if there is a $(k - 1)$-form η with $\omega = d\eta$. The subgroup of exact forms is*

$$B^k(M) = \{exact\ k\text{-}forms\ defined\ on\ M\}$$

Exact forms are analogous to boundaries in homology. Note that the integral of an exact form over a cycle is zero: Let $\omega = d(\eta)$ and let C be a k-cycle, so $\partial(C) = \emptyset$.

$$\int_C \omega = \int_C d(\eta)$$
$$= \int_{\partial(C)} \eta$$
$$= \int_\emptyset \eta$$
$$= 0$$

(11.30) Definition. *Let ω and λ be k-forms on M. Then ω is cohomologous to λ if $\omega - \lambda = d\eta$ for some $(k-1)$-form η. This is denoted by $\omega \sim \lambda$.*

Exactly as homology groups were formed, we can form cohomology groups (called cohomology since d raises instead of lowers dimensions).

(11.31) Definition. *Let M be a smooth manifold. The kth cohomology group of M, denoted by $H^k(M) = Z^k(M)/B^k(M)$, is the group of equivalence classes of elements in $Z^k(M)$ with the cohomology relation. In other words, $H^k(M)$ is $Z^k(M)$, the group of closed k-forms on M, with $\sim$ used in place of $=$.*

Appendix: Groups

(A.1) Definition. *A group is a set G with an operation $*$ defined so that if $x, y \in G$, then $x * y \in G$ and the operation $*$ satisfies the following rules:*

    (1) *Associativity: For all $x, y, z \in G$, $x * (y * z) = (x * y) * z$.*

    (2) *Identity: There is an element $e \in G$, called the identity element, such that $x * e = e * x = x$ for all $x \in G$.*

    (3) *Inverse: For each $x \in G$, there is an element $y \in G$, called the inverse of x, such that $x * y = e = y * x$.*

The group G is abelian if, in addition to the above, G also satisfies

    (4) *Commutativity: For all $x, y \in G$, $x * y = y * x$.*

The set of integers, $\mathbb{Z} = \{\ldots, -2, -1, 0, 1, 2, 3, \ldots\}$ is an abelian group under the operation of addition. The identity element is the integer 0, and the inverse of x is the integer $-x$. Similarly, the real numbers with addition, denoted $(\mathbb{R}, +)$, is an abelian group. The non-zero real numbers, $(\mathbb{R} - \{0\}, \cdot)$, is an abelian group under multiplication. The multiplicative identity is the number 1, and the inverse of x is $\frac{1}{x}$. Note that the integers $\mathbb{Z}$ with the operation of multiplication is not a group. Properties (1), (2), and (4) are true, but the integer 2 has multiplicative inverse $\frac{1}{2} \notin \mathbb{Z}$, so (3) fails.

A non-abelian group is the set of 2×2 real invertible matrices with the operation of matrix multiplication. This group is commonly denoted by

$$GL(2, \mathbb{R}) = \left\{ \begin{pmatrix} a & b \\ c & d \end{pmatrix} : a, b, c, d, \in \mathbb{R} \text{ and } ad - bc \neq 0 \right\}$$

The identity element is the identity matrix $I = \begin{pmatrix} 1 & 0 \\ 0 & 1 \end{pmatrix}$. The inverse of $A = \begin{pmatrix} a & b \\ c & d \end{pmatrix}$ is the matrix

$$A^{-1} = \frac{1}{ad - bc} \begin{pmatrix} d & -b \\ -c & a \end{pmatrix}$$

This group is not commutative, since AB is not always equal to BA. For example,

$$\begin{pmatrix} 1 & 1 \\ 0 & 1 \end{pmatrix} \cdot \begin{pmatrix} 1 & 1 \\ 1 & 0 \end{pmatrix} = \begin{pmatrix} 2 & 1 \\ 1 & 0 \end{pmatrix}$$

$$\begin{pmatrix} 1 & 1 \\ 1 & 0 \end{pmatrix} \cdot \begin{pmatrix} 1 & 1 \\ 0 & 1 \end{pmatrix} = \begin{pmatrix} 1 & 2 \\ 1 & 1 \end{pmatrix}$$

Most of our work will be with abelian groups. I usually denote the identity element by 0, instead of e, for abelian groups, and use $+$ for the operation. Property (2) of Definition A.1 implies that the identity 0 must be in every group. Let us build some examples of abelian groups.

(1) The smallest possible group is the *trivial group*, $\{0\}$, often denoted by $G = 0$, by a slight abuse of notation.

(2) If a denotes some non-zero element, a group may be formed using 0 and a. By Definition A.1(2), we must have $0 + a = a = a + 0$. We must define $a + a$. If we decide to let $a + a = 0$, then a is the inverse of a by Definition A.1(3). The set $\{0, a\}$ defined using these additions is an abelian group, denoted $\mathbb{Z}/2$.

(3) If, on the other hand, we decide that $a + a$ is not zero, let us denote $a + a = 2a$ (we are, after all, sensible people) for some convenient symbol $2a \neq 0$. Then by A.1(2), $0 + 0 = 0$, $a + 0 = a = 0 + a$, $2a + 0 = 2a = 0 + 2a$. We must choose a value for the new element $a + 2a = a + a + a = 2a + a = 3a$. We can decide that $3a = 0$, so that the inverse of a is $2a$. The set $\{0, a, 2a\}$ thus defined is denoted $\mathbb{Z}/3$.

(n) Continue thus to define $\mathbb{Z}/n = \{0, a, 2a, 3a, \ldots, (n-1)a\}$ with the convention that $na = 0$.

The groups $\mathbb{Z}/n$ defined above are called the *cyclic groups*. Note that $\mathbb{Z}/n$ has n elements.

Note that $\mathbb{Z}/4 = \{0, a, 2a, 3a\}$, with the natural operations except that $4a = 0$, is an abelian group with four elements. A different 4-element group is defined by letting a and b be two unrelated elements. The identity 0 must be in every group, and $a + 0 = a = 0 + a$, $b + 0 = b = 0 + b$. We must choose values for $a + b$, $2a = a + a$, and $2b = b + b$. If we let $2a = 0$ and $2b = 0$, then we have a new group, denoted by $\mathbb{Z}/2 \oplus \mathbb{Z}/2 = \{0, a, b, a+b\}$. We will explain this notation later.

Another way of presenting a group, besides a list of elements and rules for all the possible additions or multiplications, is to give only the *generators* and *relations*. Generators are the basic elements of the group. Every element of the group must be some combination of the generators. For example, if a group G has generators a, b, c, then typical elements of the group are called *words* and look like abc, $b^{-1}c^2ba^3$, or $abcabcabc$ in a multiplicative group. A relation on a group is a condition on the generators. Thus, if $ab = c$, then $abc = c^2$, etc. If the group is abelian, we can further reduce the words by rewriting a word in additive notation and using commutativity. A *presentation* for a group G is a list of all generators and relations. We use the notation

$$G = [[\text{ generators } : \text{ relations }]]$$
$$= [[a_1, a_2, \ldots, a_n : r_1, r_2, \ldots, r_m]]$$

for a presentation of an abelian group. For example, $\mathbb{Z}/2 = \{0, a\}$ with $2a = 0$, as described above, has generator a and relation $2a = 0$. We write $\mathbb{Z}/2 = [[a : 2a = 0]]$. Similarly, $\mathbb{Z}/n = [[a : na = 0]]$, $\mathbb{Z} = [[a]]$, and $\mathbb{Z}/2 \oplus \mathbb{Z}/2 = [[a, b : 2a = 0, 2b = 0]]$.

When presenting a group by generators and relations, one assumes that all multiples of each generator a (including $0 \cdot a = 0$ and $-1 \cdot a = -a$) are in the group, and that all additions obey the obvious rules and any exceptions are listed in the relations. Thus, $[[a]] = \{\ldots, -3a, -2a, -a, 0, a, 2a, 3a, \ldots\}$ looks just like $\{\ldots, -3, -2, -1, 0, 1, 2, 3, \ldots\} = \mathbb{Z}$. A group is *finitely generated* if it has a finite number of generators and relations.

(A.2) Definition. *A subgroup H of a group G is a subset $H \subseteq G$ such that H is also a group using the addition operation defined for G.*

Consider the group $\{e\}$ (or 0 in an additive group) consisting only of the identity element. This is a subgroup of any group. Another example is $\mathbb{Z}$ as a subgroup of $\mathbb{R}$ with the addition operation.

When checking to see if a subset H is a subgroup of G, some of the properties of Definition A.1 are inherited from the group G.

(0) First check that $x * y \in H$ for all $x, y \in H$.
(1) $x * (y * z) = (x * y) * z$ is true in H since it is true in G.
(2) One must check that $e \in H$. If so, then $e * x = x * e = x$ in H automatically.
(3) For each $x \in H$, check that the inverse $y \in H$. Then $x * y = y * x = e$ in H since it is true in G.
(4) $x * y = y * x$ in H if it is true in G.

As a shortcut, one uses the following theorem:

(A.3) Theorem. *A non-empty subset H of a group G is a subgroup if and only if for any $x, y \in H$,, $x * y^{-1} \in H$, where y^{-1} denotes the inverse of y.*

▷ **Exercise A.1**   Prove Theorem A.3.

The subset $\{0, a\} = [[a : 2a = 0]]$ is a subgroup of the group $\mathbb{Z}/2 \oplus \mathbb{Z}/2 = [[a, b : 2a = 0, 2b = 0]]$. Other subgroups of this group are $\{0\}$, $\{0, b\}$, and $\{0, a + b\}$.

▷ **Exercise A.2**   List all subgroups of $\mathbb{Z}/6$.

(A.4) Definition. *If G_1 and G_2 are abelian groups with presentations*

$$G_1 = [[a_1, a_2, \ldots, a_n : r_1, r_2, \ldots, r_m]]$$
$$G_2 = [[b_1, b_2, \ldots, b_k : s_1, s_2, \ldots, s_\ell]]$$

then the direct sum of G_1 and G_2, written $G = G_1 \oplus G_2$, is the abelian group

$$G = G_1 \oplus G_2 = [[a_1, \ldots, a_n, b_1, \ldots, b_k : r_1, \ldots, r_m, s_1, \ldots, s_\ell]]$$

Thus, $G_1 \oplus G_2$ combines all the generators and relations of both groups. The group $G = [[a, b : 2a = 0, 2b = 0]]$ discussed above is an example of a direct sum. Let $G_1 = \{0, a\} = [[a : 2a = 0]] = \mathbb{Z}/2$. Similarly, let $G_2 = \{0, b\} = [[b : 2b = 0]] = \mathbb{Z}/2$, where we choose b to represent the generator of this copy of $\mathbb{Z}/2$ to keep it distinct from the generator of G_1. The choice of the letter is, after all, arbitrary. Then G is the direct sum of G_1 and G_2, and the notation $G = \mathbb{Z}/2 \oplus \mathbb{Z}/2$ is justified.

Combining two copies of the integers gives

$$\mathbb{Z} \oplus \mathbb{Z} = [[a, b]] = \{na + mb : n, m \text{ integers}\}$$
$$= \{\ldots, -2a, -a, 0, a, 2a, \ldots, -2b, -b, b, 2b, \ldots, a + b, a - b, 2a + b,$$
$$\ldots, 2a - 47b, -72a + 37b, \text{etc.}\}$$

The a's denote one copy of the integers and the b's another. Another example is

$$\mathbb{Z} \oplus \mathbb{Z}/2 = [[a, b : 2b = 0]]$$
$$= \{\ldots, -2a, -a, 0, a, 2a, \ldots, b, a + b, -2a + b, \ldots, na + b, \ldots\}$$

▷ **Exercise A.3**   Describe the groups $\mathbb{Z}/2 \oplus \mathbb{Z}/3$, $\mathbb{Z}/2 \oplus \mathbb{Z}/2 \oplus \mathbb{Z}/2$, $\mathbb{Z} \oplus \mathbb{Z}/3$.

Some additional concepts can be adapted from linear algebra.

(A.5) Definition. *Let G be an abelian group.*

(1) *A subset $A \subseteq G$ spans G if some non-zero multiple of every element $x \in G$ can be written as a sum of multiples of the elements of A. That is, if $A = \{x_1, x_2, \ldots, x_k\}$, then every $x \in G$ satisfies*

$$nx = n_1 x_1 + n_2 x_2 + \cdots + n_k x_k$$

for some choice of integers $n, n_1, n_2, \ldots, n_k \in \mathbb{Z}$ with $n \neq 0$.

(2) *A set $A \subseteq G$ is independent if no element of A can be written as a combination of the other elements of A. Another way of saying this is set A is independent if whenever the equation*

$$0 = n_1 x_1 + n_2 x_2 + \cdots + n_k x_k$$

is true, then $n_1 = n_2 = \cdots = n_k = 0$.

(3) *A is a basis for G if A spans G and A is independent.*

Spanning guarantees that some multiple of every element of G can be a written as a sum of elements of A, and independence guarantees that A contains no redundant elements. Thus, a basis can be considered as a minimal spanning set. We say that the trivial group 0 has no basis or a basis with no elements, since $0 = 1 \cdot 0$ implies that 0 is not independent of itself. A basis is almost, but not quite, the same as a set of generators. The group $\mathbb{Z} \simeq [[a]]$ has basis $\{a\}$ since every element is of the form $n \cdot a$ for some $n \in \mathbb{Z}$. The group $\mathbb{Z} \oplus \mathbb{Z} = [[a, b]]$ has basis $\{a, b\}$ since every element $c \in \mathbb{Z} \oplus \mathbb{Z}$ can be written $c = na + mb$ for $n, m \in \mathbb{Z}$. The difference between a basis and the generators is shown in groups such as $\mathbb{Z}/2 = [[a : 2a = 0]]$. Since $2a = 0$, a is not independent of 0. Thus, the basis for $\mathbb{Z}/2$ must be the same as the basis for 0, so the basis has no elements. Similarly, $\mathbb{Z} \oplus \mathbb{Z}/2 = [[a, b : 2b = 0]]$ has basis $\{a\}$ since $2b = 0 = 0 \cdot a$, so b is in the span of $\{a\}$.

(A.6) Definition. *The rank of a finitely generated abelian group is*

$$rk(G) = \#(elements\ in\ a\ basis)$$

That this makes sense and does not depend on the particular basis chosen is proved in:

(A.7) Theorem. *G is a finitely generated abelian group. Then any two bases for G have the same number of elements.*

▷ **Exercise A.4**   Prove Theorem A.7.

Thus, $\{0\}$ has rank 0 and $\mathbb{Z}$ has rank 1. Similarly, $rk(\mathbb{Z}/2) = 0$, $rk(\mathbb{Z} \oplus \mathbb{Z}) = 2$, $rk(\mathbb{Z} \oplus \mathbb{Z}/2) = 1$.

(A.8) Theorem. *Let G_1 and G_2 be finitely generated abelian groups.*

$$rk(G_1 \oplus G_2) = rk(G_1) + rk(G_2)$$

▷ **Exercise A.5**   Prove Theorem A.8.

(A.9) Definition. *G is a abelian group with a finitely generated subgroup H. Let $x, y \in G$. Then $x \sim y \pmod{H}$ if $x - y \in H$ (that is, $x - y = h$ for some $h \in H$). The relation $\sim \pmod{H}$ is an equivalence relation. The* group formed by the elements of G with $\sim$ (mod H) used in place of $=$ is a new group denoted G/H. Elements of G/H are written either as $[x]$ or as x (mod H), where $[x] = x$ (mod$H$) denotes the set of all elements equivalent to x and is called the equivalence class of x.

An example of this process is the group $\mathbb{Z}/2$ which is really shorthand for $\mathbb{Z}/2\mathbb{Z}$. Note that $2\mathbb{Z}$ is the set of even integers and is easily seen to be a subgroup of the group of integers $\mathbb{Z}$. The new group $\mathbb{Z}/2$ is formed when $x \sim y$ (mod 2) if $x - y \in 2\mathbb{Z}$, i.e., if $x - y$ is even. Thus, $-1 \sim 1 \sim 3 \sim 5$, etc., and $-2 \sim 0 \sim 2 \sim 4 \sim \ldots$. Up to equivalence there are only two elements, so $\mathbb{Z}/2$ is the two-element group.

(A.10) Theorem. Let G be a finitely generated abelian group.

$$rk(G/H) = rk(G) - rk(H)$$

Outline of proof. Let $\{x_1, x_2, ..., x_k\}$ be a basis for H, so that $rk(H) = k$. A basis for G can be constructed from this basis for H, so the basis for G is

$$\{x_1, x_2, ..., x_k, x_{k+1}, ..., x_n\}$$

where $rk(G) = n$. It can be shown that $\{x_{k+1}, x_{k+2}, ..., x_n\}$ is a basis for G/H. ☐

(A.11) Definition. Let G_1 and G_2 be groups. A homomorphism is a function $f : G_1 \longrightarrow G_2$ such that $f(x * y) = f(x) * f(y)$ for all $x, y \in G_1$. An isomorphism is an invertible homomorphism. Groups are isomorphic, written $G_1 \simeq G_2$, if there is an isomorphism between them.

If G is any group, the *identity function*, denoted by 1_G and defined by $1_G(x) = x$ for all $x \in G$, is easily seen to be a homomorphism. Another homomorphism is the *zero or trivial homomorphism*, denoted by 0, defined for any groups G and H, by $0(x) = e$ for all $x \in G$, where e is the identity element in H.

▷ **Exercise A.6** If $f : G_1 \longrightarrow G_2$ is a homomorphism, from group G_1 to group G_2 and e_i is the identity element in G_i, prove that $f(e_1) = e_2$ and $f(x^{-1}) = (f(x))^{-1}$.

▷ **Exercise A.7** If G is a finitely generated abelian group with a finitely generated subgroup H, there is a natural homomorphism $f : G \longrightarrow G/H$ defined for $x \in G$ by $f(x) = x$ (mod H). Show that this is a well-defined homomorphism.

▷ **Exercise A.8** If G is a finitely generated abelian group with a finitely generated subgroup H, show that $G/0 \simeq G$, and $G/G \simeq 0$.

(A.12) Definition. *Let G_1 and G_2 be groups with $f : G_1 \longrightarrow G_2$ a homomorphism. The kernel of f is*

$$ker(f) = \{x \in G_1 : f(x) = 0\}$$

The range or image of f is

$$R(f) = \{y \in G_2 : y = f(x) \text{ for some } x \in G_1\}$$

▷ **Exercise A.9** Show that $ker(f)$ is a subgroup of G_1 and $R(f)$ is a subgroup of G_2.

(A.13) Theorem. *Let $f : G_1 \longrightarrow G_2$ be a homomorphism of the groups G_1 and G_2. Then*

$$rk(ker(f)) + rk(R(f)) = rk(G_1)$$

▷ **Exercise A.10** Prove Theorem A.13.

A combination of the theorems above may be used to prove the useful theorem:

(A.14) Theorem. *Let G be a finitely generated abelian group. Then G is isomorphic to a direct sum of the form*

$$G \simeq \mathbb{Z}^n \oplus \mathbb{Z}/m_1 \oplus \mathbb{Z}/m_2 \oplus \cdots \oplus \mathbb{Z}/m_k$$

where $\mathbb{Z}^n$ denotes $\mathbb{Z} \oplus \mathbb{Z} \oplus \cdots \oplus \mathbb{Z}$, the direct sum of n $\mathbb{Z}$'s, and $\mathbb{Z}/m_j$ denotes the cyclic group $[[a : m_j a = 0]]$ for $m_j \in \mathbb{Z}$. Thus, $rk(G) = n$.

Proof. Omitted.

This theorem is the most practical way of computing ranks.

▷ **Exercise A.11** Reduce the following groups to the standard form of Theorem A.14 and find their ranks.

(1) $G = [[a, b, c, d, e, f, g, h, i, j, k, \ell : a - h = 0, d + e = 0, e + f = 0,$
$b - h + i = 0, i - j - k = 0, f + g - \ell = 0, k - \ell = 0, a - g = 0,$
$j = 0, b + c = 0, c + d = 0]]$
(2) $G' = [[a, b, c, d, e, f, g, h : a - e = 0, f + g = 0, h - f - e = 0,$
$b + h = 0, c + d = 0, b - h = 0, c - 2d = 0]]$

If $f : G_1 \longrightarrow G_2$ is a homomorphism and one knows what f does to each generator of G_1, then one can calculate what f does to any element of G_1. For example, if G_1 is an abelian group generated by a, b, c, then any element of G_1 can be represented as $ka + \ell b + mc$ for some integers $k, \ell, m \in \mathbb{Z}$. Then

$$f(ka + \ell b + mc) = kf(a) + \ell f(b) + mf(c)$$

There are only two possible homomorphisms $f : \mathbb{Z} \longrightarrow \mathbb{Z}/2$, though there are many functions from $\mathbb{Z}$ to $\mathbb{Z}/2$. Let $\mathbb{Z} = \{\ldots, -2, -1, 0, 1, 2, \ldots\}$ which is generated by the number 1, and $\mathbb{Z}/2 = \{0, a\}$. One homomorphism is the trivial homomorphism defined by $0(n) = 0$ for each n. Another homomorphism is defined by letting $f(1) = a$. Since $2a = 0$ in $\mathbb{Z}/2$, $f(2) = 2f(1) = 2a = 0$, $f(3) = 3f(1) = 3a = 2a + a = 0 + a = a$, etc. Thus,

$$f(n) = \begin{cases} 0 & \text{if n is even} \\ a & \text{if n is odd} \end{cases}$$

Consider homomorphisms $f : \mathbb{Z}/2 \longrightarrow \mathbb{Z}$. We must choose $f(a)$. If $f(a) = n \in \mathbb{Z}$, then $f(2a) = f(0) = 0$, since $2a = 0$ in $\mathbb{Z}/2$. On the other hand, $f(2a) = 2f(a) = 2n \in \mathbb{Z}$. Thus, $2n = 0$ for $n \in \mathbb{Z}$. The only integer n with $2n = 0$ is $n = 0$. The only homomorphism $f : \mathbb{Z}/2 \longrightarrow \mathbb{Z}$ is the trivial one.

▷ **Exercise A.12**   List all possible homomorphisms:

(1) $f : \mathbb{Z} \longrightarrow \mathbb{Z}/3$
(2) $f : \mathbb{Z}/3 \longrightarrow \mathbb{Z}$
(3) $f : \mathbb{Z}/2 \longrightarrow \mathbb{Z}/6$
(4) $f : \mathbb{Z}/6 \longrightarrow \mathbb{Z}/2$
(5) $f : \mathbb{Z} \longrightarrow \mathbb{Z}$

Another way of representing homomorphisms for abelian groups is borrowed from linear algebra. Let $f : G \longrightarrow H$ be a homomorphism. Coordinates can be used to represent elements of G. For example, if $G = [[a, b, c : 2b = 0]]$, then the ordered triple $(1, 3, -2)$ stands for $1a + 3b - 2c$. To use this system, the set of generators and the order in which they will be referenced must be clearly stated. Also note that if the group has relations, this representation will not be unique. Above, $(1, 3, -2) = a + 3b - 2c = a + b + 2b - 2c = a + b + 0 - 2c = a + b - 2c = (1, 1, -2)$. If the elements of G and H are represented in coordinate form, then the homomorphism f can be represented by a matrix. To form a matrix representing $f : G \longrightarrow H$, first fix generating sets $x_1, x_2, \ldots, x_n$ for G and $y_1, y_2, \ldots, y_m$ for H, then take each generator x_i of G, in order, and figure out $f(x_i)$. Write $f(x_i)$ in vector form as a combination of the y_j's in H.

The $m \times n$ matrix for f is then formed by setting $f(x_i)$ in the matrix as the ith column.

$$f = \left[f(x_1) \mid f(x_2) \mid \cdots \mid f(x_n) \right]$$

Multiplying a vector or ordered n-tuple representing an element x of G by the matrix gives another vector, which will represent $f(x)$ in H.

▷ **Exercise A.13** Consider the homomorphism

$$f : \mathbb{Z} \oplus \mathbb{Z} \oplus \mathbb{Z}/2 \oplus \mathbb{Z} \oplus \mathbb{Z}/2 \longrightarrow \mathbb{Z} \oplus \mathbb{Z}/6 \oplus \mathbb{Z}/3 \oplus \mathbb{Z}/2$$

defined by $f(a) = w, f(b) = w, f(c) = 3x, f(d) = 0, f(e) = z$, where

$$\mathbb{Z} \oplus \mathbb{Z} \oplus \mathbb{Z}/2 \oplus \mathbb{Z} \oplus \mathbb{Z}/2 \simeq [[a, b, c, d, e : 2c = 0, 2e = 0]]$$

$$\mathbb{Z} \oplus \mathbb{Z}/6 \oplus \mathbb{Z}/3 \oplus \mathbb{Z}/2 \simeq [[w, x, y, z : 6x = 0, 3y = 0, 2z = 0]]$$

(1) Find a matrix representing f.
(2) Using your answer from (1), compute the following and then rewrite as a combination of the generators:
    (i) $f(a + b + c + d + e)$
    (ii) $f(2a + 3b + 2c - 4d - e)$

References

These texts are of the same difficulty level as this book.

D. W. Blackett, *Elementary Topology*, Academic Press, 1982.
F. Croom, *Principles of Topology*, Saunders, 1989.
V. Guillemin and A. Pollack, *Differential Topology*, Prentice-Hall, 1974.
M. Henle, *A Combinatorial Introduction to Topology*, W. H. Freeman, 1979.
C. Kosniowski, *A First Course in Algebraic Topology*, Cambridge University Press, 1980.
E. E. Moise, *Geometric Topology in Dimensions 2 and 3*, Springer-Verlag, 1977.
E. O. Roxin, *Ordinary Differential Equations*, Wadsworth, 1972.
T. L. Saaty and P. C. Kainen, *The Four-Color Problem*, McGraw-Hill, 1977.
J. Stillwell, *Geometry of Surfaces*, Springer-Verlag, 1992.
C. T. C. Wall, *A Geometric Introduction to Topology*, Addison-Wesley, 1972.
J. R. Weeks, *The Shape of Space*, Marcel Dekker, 1985.

These are texts at a more advanced level.

G. E. Cooke and R. L. Finney, *Homology of Cell Complexes*, Princeton University Press, 1967.
A. A. Kosinski, *Differentiable Manifolds*, Academic Press, 1993.
W. Massey, *A Basic Course in Algebraic Topology*, Springer-Verlag, 1991.
J. R. Munkres, *Topology, a First Course*, Prentice-Hall, 1975.
C. P. Rourke and B. J. Sanderson, *Introduction to Piece-wise Linear Topology*, Springer-Verlag, 1982.
H. Seifert and W. Threlfall, *A Textbook of Topology*, Academic Press, 1980.
A. J. Sieradski, *An Introduction to Topology and Homotopy*, PWS-Kent, 1992.
I. M. Singer and J. A. Thorpe, *Lecture Notes on Elementary Topology and Geometry* Springer-Verlag, 1967.
J. Stillwell, *Classical Topology and Combinatorial Group Theory*, Springer-Verlag, 1980.
J. W. Vick, *Homology Theory*, Academic Press, 1973.

Here is a sampling of articles and booklets on topology.

R. H. Bing, *The elusive fixed point property*, Amer. Math. Monthly **76** (1969) 119–132.

M. H. Brill and M. Stueben, *Demystifying the projective plane*, Math. Mag. **63** (1990) 48–51.

K. C. Cole, *Escape from 3-D*, Discover **14** (1993) 52–62.

A. Durfee, N. Kronenfeld, H. Munson, J. Roy, and I. Westby, *Counting critical points of real polynomials in two variables*, Amer. Math. Monthly **100** (1993) 255–271.

J. H. Fife, *The Kuratowski closure-complement problem*, Math. Mag **64** (1991) 180–182.

V. F. R. Jones, *Knot theory and statistical mechanics*, Sci. Amer. **263** (1990) 98–103.

J. Milnor, *Topology from the Differentiable Viewpoint*, University Press of Virginia, 1965.

T. L. Moore, *Using Euler's formula to solve plane separation problems*, Coll. Math J. **22** (1991) 125–130.

L. Neuwirth, *The theory of knots*, Sci. Amer. **240** (1979) 110–124.

Yu. A. Shashkin, *Fixed Points*, American Mathematical Society and Mathematical Association of America, 1991.

G. Schwarz, *The dark side of the Moebius strip*, A. M. Monthly, **97** (1990) 890–897.

W. Tape, *The topology of mirages*, Sci. Amer. **252** (1985) 120–129.

G. Taubes, *What happens when hubris meets nemesis*, Discover **8** (1987) 66–77.

C. Thomasson, *The Jordan-Schönflies theorem and the classification of surfaces*, Amer. Math. Monthly **99** (1992) 116–130.

W. Thurston and J. R. Weeks, *The mathematics of three-dimensional manifolds*, Sci. Amer. **251** (1984) 108–120.

T. Yung Kong, R. Kopperman, P. R. Meyer, *A topological approach to digital topology*, Amer. Math. Monthly **98** (1991) 901–917.

Index

Undergraduate Texts in Mathematics

Learning Resources
Centre